A Guide to Wider Horizons

Kevin Krisciunas

George P. and Cynthia Woods Mitchell Institute for Fundamental Physics and Astronomy
Texas A&M University

Cover image credit: NASA: A view of the horizon from the International Space Station that allows scientists to detect the gases and particles that make up the different layers of our atmosphere. Astronauts aboard the International Space Station captured this digital photograph on July 31, 2011.

www.kendallhunt.com
Send all inquiries to:
4050 Westmark Drive
Dubuque, IA 52004-1840

Copyright © 2014 by Kevin Krisciunas

ISBN 978-1-4652-3894-8

Kendall Hunt Publishing Company has the exclusive rights to reproduce this work, to prepare derivative works from this work, to publicly distribute this work, to publicly perform this work and to publicly display this work.

All rights reserved. No part of this publication may be reproduced, stored in a retrieval system, or transmitted, in any form or by any means, electronic, mechanical, photocopying, recording, or otherwise, without the prior written permission of the copyright owner.

Printed in the United States of America
10 9 8 7 6 5 4 3 2 1

Contents

Acknowledgments v

1 Introduction 1
2 Women Runners are $11\frac{1}{2}$ Percent Slower than Men 5
3 The Energy Budget 9
    Kinetic Energy and Potential Energy   9
    The World Record in the Pole Vault   10
    Collision of a Small Asteroid with the Earth   12
    The Lifetime of the Sun   13
4 Some Curious Words, Many of Which We Know 15
5 The Good, the Bad, and the Obsessive 21
    The Most Brilliant Ph.D. Thesis Ever Written in Astronomy   21
    The Most Egotistical Book in the History of Astronomy   22
    Not for the Thin-Skinned   25
6 How Long Do Astronomers Live? 29
7 Astronomical Measurements without a Telescope 37
8 Improvements in Astronomical Imaging 43
9 The Constellations 49
10 A Mathematical Sampler 53
    Integration for People Who Don't Know Calculus   53
    Drawing Figures with a Ruler and Compass   55
    Some Fun with Angles   56

Three Mathematical Geniuses   57
   Ramanujan   57
   Euler   59
   Gauss   61

11 Homework 67
12 A Breed Apart 73
13 How to Get People to Tell the Truth............................ 79
14 Famous First Words 83
15 Books with a Moral Angle 89
16 Lost Books 95
17 A Novel Concept 99
18 Recommended Nonfiction 105
   Asking Some of the Right Questions   105
   The Moral Animal   107
   The Wealth and Poverty of Nations   109
   Can't We All Get Along?   110
   Final Recommendations   111

Epilogue: The Importance of the Trial of Galileo 115
Index 119

Acknowledgments

I thank the following people for their comments on various sections of this book: Rebecca Shaftman, Tim Abbott, Shana Loshbaugh, Candace Campanelli, Bill Bassichis, and Saleh Al-Ghusson. I also thank Judy Lola Bausch, Amanda Ritzman, Owen Gingerich, Dieter Herrmann, Ian Gatley, Sam Gatley, Michael Rynders, and my father, Alfonse Krisciunas, for references and insights. Jude Magaro, Peter Nugent, and Nick Suntzeff made key suggestions for the chapter called "Homework." I thank Ian Ridpath, Robert Zaretsky, and Robert Blake, Jr., for permission to quote their work. Blake's winning entry to the Faux Faulkner contest of 1998 is reproduced courtesy of Yoknapatawpha Press.

1

Introduction

This is a book about books and a book about ideas. It is a book about places to begin and some curious and weird things that are probably all true. It is a book about finding a good book to read. It is a book about branching out into new areas, about learning some things about literature, history, science, and mathematics. It is about enriching your life, about widening your horizons, and perhaps about adding some spice to your conversation. If any of the jumping-off spots cause you to accept some of my recommendations, then I will have succeeded.

Prior to the invention of movable type by Johannes Gutenberg (about 1439), books were very rare commodities because they had to be copied by hand, one at a time. Printing allowed many identical copies to be made, but printing runs rarely exceeded a few hundred copies. Still, knowledge has been disseminated at a more and more rapid rate up to the present day.

From 1751 to 1772, French, Prussian, and British scholars published the *Encylopédie*. It was an attempt to summarize everything that was known at the time concerning science, arts, and crafts. Today the US Library of Congress contains over thirty-two million books in 470 languages, and there are billions of web pages. Now, if you read a book a week, that amounts to fifty-two books in a year and twenty-six hundred books over fifty years. A lifetime of reading by an avid reader could pretty easily fit in the average American house. Even the best of us can only scratch the surface of all that has been published. Where should a humble person, eager to learn, begin?

I would not have an interest in reading a whole volume of any general encyclopedia, but I once read the 102 chapters of *The Columbia History of the World* (1987), which runs to 1237 pages. Writer David Denby went back to Columbia University at the age of forty-eight and retook Western Civilization. Then he wrote a lovely book called *Great Books* (1997), which is meatier than Cliff Notes but not as demanding as many volumes of the Great Books of the Western World championed by the University of Chicago.

1 Introduction

Figure 1.1 The author talking about Renaissance astronomy. *Photo by Shana Hutchins.*

Carl Sagan said:

> What an astonishing thing a book is. It is a flat object made from a tree with flexible parts, on which are imprinted lots of funny dark squiggles. But one glance at it and you're inside the mind of another person. [...] Writing is perhaps the greatest of human inventions, binding together people who never knew each other, citizens of distant epochs. Books break the shackles of time. Books are proof that humans are capable of working magic.[1]

Now when it comes to authenticity in scholarship, there is a gray zone. Helen Schmerer, who taught reference librarianship at the University of Chicago, said, "There is no *truth*, just the facts according to some particular source." As an example, let us consult the famed eleventh edition of the *Encyclopaedia Britannica* (1910–1911), considered by some to be the most authoritative encyclopedia ever published.

This particular example is about Girolamo Savonarola (1452–1498), a rabble-rousing Dominican friar who became *de facto* ruler of Florence. He was eventually burned at the stake for his sermons against corrupt (i.e., rich) citizens and Catholic clerics, particularly the Borgia family, whose patriarch was Rodrigo Borgia, also known as Pope Alexander VI. In the eleventh edition of the *EB* we find this passage:

> Pleasure-loving Florence was completely changed. Abjuring pomps and vanities, its citizens observed the ascetic régime of the cloister; half the year was devoted to abstinence

and few dared to eat meat on the fasts ordained by Savanorola. Hymns and lauds rang in the streets that had so recently echoed with Lorenzo [de Medici's] dissolute songs. Both sexes dressed with Puritan plainness; husbands and wives quitted their homes for convents; marriage became an awful and scarcely permitted rite; mothers suckled their own babes; and persons of all ranks—nobles, scholars and artists—renounced the world to assume the Dominican robe. *Still more wonderful* [my emphasis] was Savonarola's influence over children, and their response to his appeals is a proof of the magnetic power of his goodness and purity.

"Still more wonderful"? Since when is abandoning your family and the sacrament of marriage part of the Christian heritage? Clearly, the author of this article departed from objectivity. This author was Linda Villari, wife of Pasquale Villari, who published a two-volume biography of Savonarola in 1859 and 1861.

One time while leafing through a book on interesting mathematical things I came across a result given in our chapter "A Mathematical Sampler" as Equation 10.9. It turns out that the compiler of that book did not realize the author of a particular *Scientific American* article was pulling an April Fool's Day prank on the reader. We all know that you cannot believe everything you read, but it is always surprising when you realize you have been duped.

As pointed out by Samuel Arbesman in his book *The Half-Life of Facts*, "Whatever fact first appears in print, whether true or not, is very difficult to dislodge."[2] Did you know that a Brontosaurus is really named something else (i.e., apatosaurus)?[3]

Never mind what you know about exponential growth rates. The writer who asserted nearly forty years ago that 75 percent of the people who had ever been born are alive today was *wrong*. A *lower* estimate of the number of people who have lived on Earth since 50,000 BC is about 108 billion! Only about 6.5 percent of all people ever born are alive today.[4]

As you read the material presented here, ask yourself the following question: "How can this be true or accurate?" Some of it is pretty bizarre, but so far as I can tell, it is all true, or at least it is according to the sources referenced.

Endnotes

1. http://writersalmanac.publicradio.org/index.php?date=2012/11/09 (accessed December 31, 2012).
2. Arbesman, Samuel, *The Half-Life of Facts: Why Everything We Know Has an Expiration Date*, New York: Current, 2012, p. 86.
3. Ibid., p. 79.
4. http://www.prb.org/Articles/2002/HowManyPeopleHaveEverLivedonEarth.aspx (accessed February 14, 2013).

2

Women Runners are $11\frac{1}{2}$ Percent Slower than Men

In Table 2.1 we show the world records for running, from 100 m to the marathon. The last column gives the women's best time divided by the men's best time. The median ratio is 1.119. By definition, 50 percent of the data are larger than the median and 50 percent are smaller. The median value is often more stable than the average of a small number of data points.

In Table 2.1 we also show the times of the Olympic champions from the 2012 summer Olympics. We exclude the mile (which is not run in the Olympics), but include the 4 × 100 m relay, the 4 × 400 m relay, and the Olympic triathlon, which consists of a 1500 m swim, a 40 km bike race, and a 10 km run. For these Olympic events the median ratio of women's winning time to men's winning time is 1.115. This can be compared to the results for running events (excluding relays and the triathlon) of the 1996, 2000, 2004, and 2008 Olympics, which give median ratios of women's winning time to men's winning time of 1.114, 1.115, 1.114, and 1.125.

In thinking about this since the 1996 summer Olympics, I conclude that there is something fundamental about the number 1.115. For world-class runners the women are typically 11.5 percent slower than the men. This seems to be a stable number related to the basic design of the female body compared to the male body.

Now, if you are a betting person and you witness the women's final, you can take the women's winning time, divide by 1.115, and have a very good prediction of the men's winning time. You cannot say *who* will run this, but you can predict the time accurately. Or, if the men run first, take the men's winning time and multiply by 1.115, and that is a good prediction for the women's winning time.

The Olympic running events that are the most unpredictable are the 1500 m and the 5 km races. The world records are never threatened. Everyone is just trying to win the race, and the best strategy for that is to "jog" the first three-quarters of the race then drive to the finish as hard as possible.

Table 2.1 Athletics Records

Data Set[a]	Event	Men's Time	Women's Time	Ratio
1	100 m	9.58	10.61[b]	1.108
1	200 m	19.19	21.34	1.112
1	400 m	43.18	47.60	1.102
1	800 m	1:40.19	1:53.28	1.123
1	1500 m	3:26.00	3:50.46	1.119
1	mile	3:43.13	4:12.56	1.132
1	5 km	12:37.35	14:11.15	1.124
1	10 km	26:17.53	29:31.78	1.123
1	marathon	2:03:38[c]	2:15:25	1.095
2	100 m	9.63	10.75	1.116
2	200 m	19.32	21.88	1.133
2	400 m	43.94	49.55	1.128
2	4 × 100 m	36.84	40.82	1.108
2	800 m	1:40.19	1:56.19	1.151
2	1500 m	3:34.08	4:10.23	1.161
2	4 × 400 m	2:56.72	3:16.87	1.114
2	5 km	13:41.66	15:04.25	1.101
2	10 km	27:30.42	30:20.75	1.103
2	marathon	2:08:01	2:23:07	1.118
2	triathlon	1:46:25	1:59:48	1.126
3	Hawaii Ironman	8:03:56	8:54:02	1.104

[a] 1 = World records from[1]. 2 = 2012 London Olympics winning times from NBC Sports website at the time of the Olympics, but the data can also be found in reference[2]. Times are in seconds, mm:ss.s, or hh:mm:ss.s. 3 = Hawaii Ironman course records.[3]
[b] Official women's WR is 10.49, but wind gauge was not working and wind was above legal limit.
[c] 2011 Boston marathon was actually faster, but course was downhill point to point and wind aided.

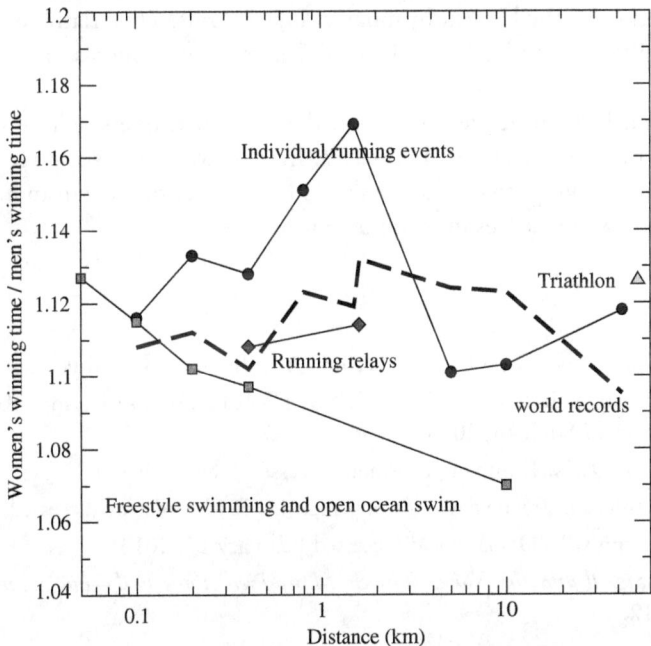

Figure 2.1 Ratio of women's winning time to men's winning time from 2012 summer Olympics. Running data are from Table 2.1. Dashed line is based on world records.

How about amateur runners? Let us consider the Bryan/College Station half marathon, which was run on December 9, 2012, where I live in Texas.[4] There were 779 male finishers and 1310 female finishers. The men's winner ran 1:16:08. The women's winner ran 1:24:41. The women's winning time divided by the men's winning time gives a ratio of 1.112. The tenth-place finishers give a ratio of 1.103. The top tenth percentile of each gender gives a ratio of 1.155. The top twenty-fifth percentile gives a ratio of 1.131. The fiftieth percentile gives a ratio of 1.123. That the women were 12 percent slower is essentially the same result as for the world records.

How about swimmers? In the 2012 summer Olympics we may consider the 50 m, 100 m, 200 m, 400 m freestyle races, and the 10 km open ocean swim. The ratios of the women's winning time to the men's winning time were 1.127, 1.115, 1.102, 1.097, and 1.070.

Ratios of women's winning time to men's winning time are shown in Figure 2.1. In contrast to the results for running events and the Olympic triathlon, it seems that as we progress to longer and longer swims, the women get more and more competitive with the men. The ratio drops. If there were a very long open ocean swimming race, it might be won by a woman. The swimming career of Diana Nyad would be an example to follow.[5] On September 2, 2013, she became the first person to swim the 110 miles from Cuba to Key West, Florida, without a shark cage.

In Table 2.1 we also give the Hawaii Ironman course records. This triathlon involves a 2.4 mile swim, a 112 mile bike ride, and a 26.2 mile run. The ratio of women's to men's record times is 1.104.

The next time you hear someone say, "Soon the women runners will be beating the men," you should just let the data speak for themselves. Even as women's world records and Olympic performances keep improving, so do the men's. As Nate Silver points out in his book *The Signal and the Noise*,[6] the data themselves must guide our opinions.

Endnotes

1. http://en.wikipedia.org/wiki/List_of_world_records_in_athletics (accessed February 3, 2013).
2. http://www.flotrack.org/article/13720-FULL-RESULTS-2012-London-Olympic-Games-Track-and-FieldAthletics (accessed March 14, 2013).
3. http://en.wikipedia.org/wiki/Ironman_Triathlon (accessed July 18, 2013).
4. http://onlineraceresults.com/race/view_race.php?race_id=30133#racetop (accessed September 9, 2013).
5. http://en.wikipedia.org/wiki/Diana_Nyad (accessed February 15, 2013).
6. Silver, Nate, *The Signal and the Noise: Why So Many Predictions Fail—but Some Don't*, New York: Penguin Press, 2012.

3

The Energy Budget

Kinetic Energy and Potential Energy

Consider a bowling ball of mass m placed on a closet shelf two meters high. We have given it potential energy (PE) relative to the floor equal to

$$\text{PE} = mgh, \tag{3.1}$$

where $h = 2$ m and g is the local acceleration of gravity, 9.8 m/sec². If the mass of the bowling ball is 7 kg, the potential energy is 137.2 Joules (abbreviated J).

Now say a train passes nearby and the slight vibrations in the house cause the bowling ball to roll off the shelf and fall to the floor. In the absence of air friction, as it falls its speed will increase by 9.8 m/sec each second.

Galileo discovered that an object accelerating uniformly with acceleration a will traverse a distance

$$d = \frac{1}{2}at^2. \tag{3.2}$$

So our bowling ball will fall to the floor in $t = \sqrt{2h/g} = 0.6389$ sec. Its velocity when it hits the floor will be

$$v = gt = 9.8 \text{ m/sec}^2 \times 0.6389 \text{ sec} = 6.261 \text{ m/sec}. \tag{3.3}$$

Its *energy of motion* (or *kinetic energy*) when it hits the floor will be

$$\text{KE} = \frac{1}{2}mv^2, \tag{3.4}$$

or $0.5 \times 7 \text{ kg} \times (6.261 \text{ m/sec})^2 = 137.2$ J. Note that the kinetic energy when the bowling ball hits the floor is exactly the same as the potential energy when the bowling ball is on the shelf two meters above the floor. As the bowling ball falls, potential energy is converted into kinetic energy. The sum of the two at any point during the fall equals a constant, which we call the total energy. For such a situation

$$\text{KE} + \text{PE} = \text{total energy} = \text{a constant.} \qquad (3.5)$$

The only stipulation here is that there are no dissipative forces such as friction. Equation 3.5 also holds for the case of a planet orbiting the Sun on an elliptical orbit. Its speed and distance will vary, but the total orbital energy will be some constant. In the case of a mass attached to a spring and sliding back and forth on a horizontal surface, Equation 3.5 will also hold, though there is a different expression for the potential energy.

The World Record in the Pole Vault

Now let us consider the case of a sprinter running one hundred meters. He is down in the blocks and the gun goes off. It takes him a while to react to the gun and get moving. He is actually moving forward about 0.3 seconds after the gun. He accelerates to about 80 percent of his maximum speed over the next second, then from 80 percent to maximum speed over the next two seconds. Then he runs at his maximum speed to the finish line. This is illustrated in Figure 3.1. Another way of saying this is that a sprinter reaches his maximum speed about twenty-five meters from the start. Pole vaulters use an approach a little longer than this to achieve maximum speed.

Since velocity is the rate of change of position, if we plot the velocity versus time, the area under the velocity curve from time $t = 0$ to $t = t_{end}$ is equal to the distance covered. But for our

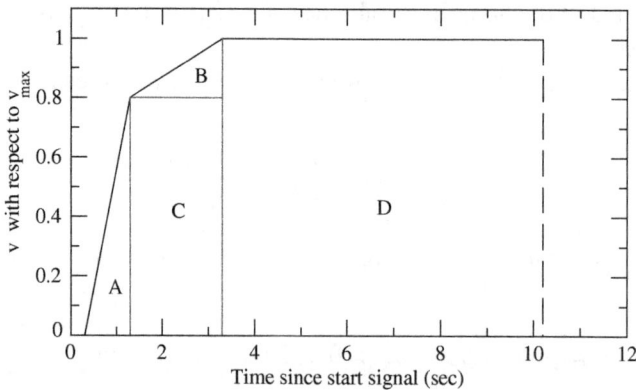

Figure 3.1 Velocity compared to maximum velocity for a sprinter who can run the hundred-meter dash in 10.2 seconds. The area under the solid black line gives the distance traveled, which is one hundred meters.

sprinter just described, we know the distance covered. That is one hundred meters. To use some phraseology from calculus, the integral of the velocity gives us the distance.

Given our model of how a sprinter's velocity varies from the gun to the finish line, we can estimate the maximum velocity by first computing the area under the curve in Figure 3.1. Area A is equal to 0.4 v_{max}. Area B is equal to 0.2 v_{max}. Area C is equal to 1.6 v_{max}. And area D is equal to $(t_{end} - 3.3)v_{max}$. Altogether we have:

$$(t_{end} - 3.3)v_{max} + 2.2v_{max} = 100 \text{ m}. \tag{3.6}$$

The world record in the pole vault (set in 1994 outdoors, 1993 indoors) is held by the Ukrainian athlete Sergei Bubka. Let me take an educated guess that in his prime Bubka could have run one hundred meters in 10.2 sec. This is only about 3.5 percent slower than the world record at that time. Bubka was the fastest vaulter, which allowed him to hold higher on the vaulting pole and use a stiffer pole than other vaulters.

Using Equation 3.6 we obtain v_{max} = 10.99 m/sec for Bubka.[1,2] Next, if Bubka could have converted his kinetic energy with 100 percent efficiency into potential energy, we could equate

$$\frac{1}{2}mv_{max}^2 = mgh_{max}. \tag{3.7}$$

Notice that we have mass on both sides of the equation, so it divides out, leaving us with

$$h_{max} = \frac{v_{max}^2}{2g}. \tag{3.8}$$

With v_{max} = 10.99 m/sec and g = 9.8 m/sec^2, h_{max} = 6.16 m. This is to be compared with Bubka's actual world indoor record of 6.15 meters, and his outdoor record of 6.14 meters. The agreement is remarkable! Thus, a simple consideration of kinetic energy, potential energy, and an algorithm for determining a sprinter's maximum velocity allows a prediction that matches reality very closely.

In the previous chapter we found that from sprints to long distance, women on average run 1.115 times slower than men. The women's world record in the pole vault is 5.06 meters, by Yelena Isinbayeva in 2009. If her maximum speed was 10.99/1.115 = 9.86 m/sec, then Equation 3.8 predicts that her maximum vault should be 4.96 meters. She exceeded this by ten centimeters (about four inches).

But we have swept some complications under the rug. A vaulter runs more slowly with a pole than without one. And the change in height should be the change of the *center of mass*. From reference[1] we have an estimate that Bubka's speed was 9.9 m/sec as he went down the runway. Equation 3.8 would predict a change in height of his center of mass of 5.00 meters. The average ratio of the center of mass to height is 0.560 for men and 0.543 for women.[3] Bubka was 1.83 meters tall, so his center of mass was roughly 1.025 meters above ground. A prediction of his best vault would be 5.00 + 1.025 = 6.025 m. This implies that his center of mass went under the bar by 11.5 to 12.5 cm for his world record jumps.

For the women's world record holder (1.73 meters tall), her center of mass was about 0.95 meters above ground. Her maximum speed can be estimated to be $9.9/1.115 = 8.88$ m/sec. Equation 3.8 gives a vertical change of her center of mass of 4.02 meters, so a prediction of her best vault would be $4.02 + 0.95 = 4.97$ m. She achieved 5.06 meters, implying that her center of mass went under the bar by 9 centimeters. The center of mass of a world-class high jumper or pole vaulter goes under the bar, even though his or her body goes over the bar.

Collision of a Small Asteroid with the Earth

Rocks from outer space collide with the Earth every day. An asteroid the size of a town might collide with the Earth once every hundred million years. Let us calculate the kinetic energy of such a collision.

Imagine a spherical asteroid five kilometers in diameter (3.1 miles). What is its volume? The volume of a sphere is $\frac{4}{3}\pi r^3$. With $r = 2500$ meters, the volume is 6.55×10^{10} (65.5 billion) cubic meters.

Next, what is the mass of the asteroid? For that we must assume a density. The density of water is one gram per cubic centimeter, by definition, which is the same as one thousand kilograms per cubic meter. If you dig up a bucket of rocks and dirt in your garden, you will find that the average density of this material is about twice that of water, or 2000 kg/m^3. The Earth's average density is higher, about 5.5 times that of water, because its core is mostly made of iron and nickel, which are considerably denser than the dirt and rocks from your garden. So, assuming a rather low density of twice that of water, our hypothetical asteroid would have a mass of 1.31×10^{14} kg.

The speed of the Earth around the Sun is 30 km/sec $= 3 \times 10^4$ m/sec. If our hypothetical asteroid had a rear-end collision with us, the relative speed would likely be less than 30 km/sec. If the asteroid came directly toward us at 20 km/sec, the net velocity of this head-on collision would be 50 km/sec. Let us just say the asteroid is crossing the Earth's orbit perpendicularly, so the effective speed of the collision is just the Earth's speed around the Sun.

Now we calculate the kinetic energy of the collision.

$$\text{KE} = \frac{1}{2}mv^2 = 0.5 \times 1.31 \times 10^{14} \times (3 \times 10^4)^2 = 5.89 \times 10^{22} \, J. \tag{3.9}$$

This is a lot of Joules. When it comes to destructive power, we like to use the equivalent of a ton of dynamite, or even a million tons (megaton). How much energy do we get from blowing up a ton of dynamite? It is 4.18×10^{15} J. The result is that our relatively small, relatively low-density asteroid has the destructive power of a fourteen *million* megaton nuclear warhead.

By contrast, consider the biggest bomb ever detonated by humans, the so-called Tsar Bomba, exploded by the Soviet Union on October 30, 1961. It had a yield of fifty-eight megatons (see Figure 3.2).[4] Thus, the potential destructive power of the collision of a small asteroid with the Earth is more than two hundred thousand times more than that of the largest bomb humans have produced!

3 The Energy Budget   13

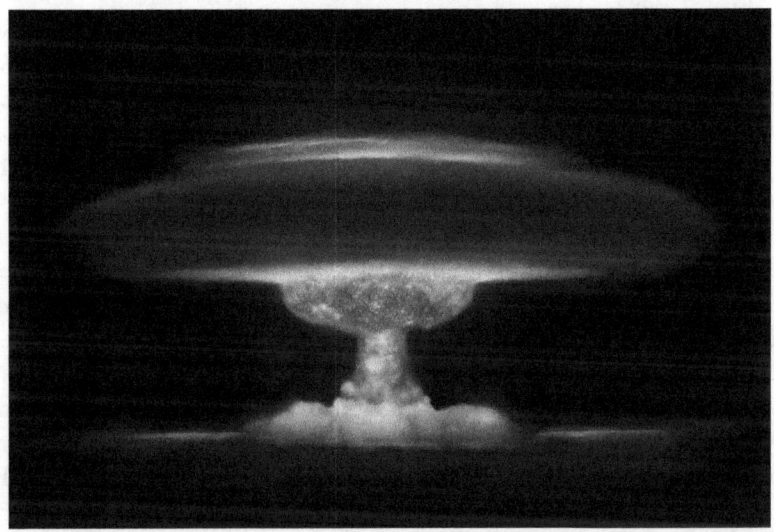

Figure 3.2 Atmospheric tests of nuclear weapons, such as this explosion, occurred many times in the 1950s and early 1960s. The Tsar Bomba was the most powerful bomb ever produced by humans. It was detonated by the Soviet Union on October 30, 1961, and had a yield equivalent to fifty-eight megatons of dynamite. *Image © Torquetum, 2013. Used under license from Shutterstock, Inc.*

The Lifetime of the Sun

The Sun is powered by nuclear fusion in its core. There, hydrogen nuclei (protons) are fused into helium nuclei via a process known as the *proton-proton* cycle.

The mass of four hydrogen nuclei is 6.693×10^{-27} kg. The mass of one helium nucleus is 6.645×10^{-27} kg. The difference is 0.048×10^{-27} kg. Thus, the fraction of mass that is converted into energy is $0.048/6.693 = 0.0072$, or 0.72 percent.

Let us naively assume that when the Sun was formed it was 100 percent hydrogen and that all of those protons will be converted into helium nuclei. How long would that take?

Most everyone has heard of Einstein's famous formula $E = mc^2$. To calculate the amount of energy (in Joules) that are obtained from the conversion of mass into energy you multiply the mass m by the square of the velocity of light c, which is 3.0×10^8 m/sec.

The mass of the Sun is 2×10^{30} kg. Thus, in our naive model we convert 0.0072 of this into energy, or $0.0072 \times 2 \times 10^{30}$ kg $= 1.44 \times 10^{28}$ kg. The amount of energy liberated via nuclear fusion will be $E = mc^2 = 1.44 \times 10^{28} \times (3 \times 10^8)^2 = 1.3 \times 10^{45}$ J. Since the Sun gives off 3.8×10^{26} Joules of energy each second it could keep this up for $1.3 \times 10^{45}/3.8 \times 10^{26} = 3.4 \times 10^{18}$ seconds. A year consists of roughly 3.156×10^7 seconds, so the time scale is 108 billion years!

It turns out that the Sun was only 75 percent hydrogen when it was formed. Also, hydrogen can only be converted to helium if the temperature is greater than ten million deg K. This is the case for the core, the inner 20 percent of the radius of the Sun. Once 10 to 15 percent of the original hydrogen of the Sun is converted into helium, the Sun will become a red giant star.

This stage of hydrogen conversion in the core is called the *main sequence* stage of the Sun's life. How long does it last? About 108 billion years $\times$ 0.75 $\times$ 0.125 = 10 billion years.

Stars more massive than the Sun have hotter core temperatures and burn their core fuel more rapidly. Counter to what you might think, they have *shorter* lifetimes. Stars less massive than the Sun burn their core fuel much more slowly and have very long lifetimes. But the bottom line is that the timescale for stellar evolution must be measured in millions or billions of years. There is no getting around it.

I acknowledge that many people believe that the Earth is six thousand years old on the basis of a chronology from the Bible. But the Bible was intended as a book of history and moral guidance. It is not a book of science. A friend of Galileo, Caesar Cardinal Baronius, said, "The Bible teaches us how to go to Heaven, not how the heavens go."[5] We should not rely on it for everything. Even St. Thomas Aquinas (1225–1274) is reported to have said, "Timio hominem unius libri." Translation: I fear the man of one book.[6]

Let us consider one particular Biblical verse, the second epistle of Peter, chapter 3, verse 8. "To the Lord a day is like a thousand years, and a thousand years is like a day." You may have your own interpretation of this verse, but here is mine. Concerning the age of the universe, God does not care about the number or the units. He only wants you to recognize that He made this place. It is our job to figure out *how* He made this place.

Endnotes

1. I note that an undocumented fact given in the Wikipedia article on Bubka is that his *average* speed during the pole vault approach was 9.9 m/sec. See http://en.wikipedia.org/wiki/Sergey_Bubka (accessed February 8, 2013).
2. At http://sports.espn.go.com/oly/summer08/fanguide/athlete?athlete=1794 (accessed February 8, 2013) we read that Bubka was capable of running one hundred meters in 10.3 sec. That would translate into a maximum speed of 10.87 m/sec and an implied pole vault potential of 6.03 m, given our simple model.
3. http://hypertextbook.com/facts/2006/centerofmass.shtml (accessed March 20, 2013).
4. http://news.bbc.co.uk/onthisday/hi/dates/stories/october/30/newsid_3666000/3666785.stm (accessed August 14, 2013).
5. Sobel, Dava, *Galileo's Daughter: A Historical Memoir of Science, Faith, and Love*, New York: Walker & Co., 1999, p. 65. Also: Gingerich, Owen, "The Galileo Affair," *Scientific American*, 247, no. 2 (August 1982), pp. 132–143, on p. 134.
6. Sarton, George, *Six Wings: Men of Science in the Renaissance*, Bloomington, Indiana: Indiana University Press, 1957, p. 61.

4

Some Curious Words, Many of Which We Know

Language is at least eighty thousand years old, but writing only began about five thousand five hundred years ago. There are about six thousand languages in the world, but only about two hundred are written regularly. Every language contains words from other languages.[1]

Richard Francis Burton (1821–1890), best known for his translation of *One Thousand and One Nights*, understood upward of thirty languages.[2] Only a handful of people in the world could make a similar claim today, though I met a graduate engineering student recently who said he knew thirty *programming* languages. While a person might say, "I only know English," there are many words from other languages that we use, often without thinking where they come from.

James Nicoll wrote, "The problem with defending the purity of the English language is that English is about as pure as a cribhouse whore. We don't just borrow words; on occasion, English has pursued other languages down alleyways to beat them unconscious and riffle their pockets for new vocabulary."[3]

Think of how many French terms we use, such as rendezvous, tête-à-tête, laissez-faire, cul-de-sac, debonnaire, ménage à trois, chic, communiqué, haute couture, détente, rapprochement, fleur-de-lis, niche, genre, route, croissant, ouevre, manoeuvre, petits fours, hors d'oeuvres, bon-bon, blasé, métier, mélange, outré, c'est la vie, entrepreneur, potpourri, baguette, echelon, cliché, attaché, esprit de corps, coup d'état, joie de vivre, carte blanche, raison d'être, avant-garde, bon vivant, bon voyage, camouflage, entourage, espionage, rapportage, sabotage, savoir faire, ennui, ingénue, tour de force, foyer, faux, naïve, reconnaissance, raconteur, débâcle, bonhomie, chauffeur, critique, façade, femme fatale, film noir, brouhaha, silhouette, soufflé, papier-mâché, savoir faire, voila!

The word *forte*, pronounced as two syllables, is Italian and means "loud." But when we say, for example, "Writing good prose is your *forte*," the word *forte* is the French version and has only one syllable (like "fort").

4 Some Curious Words, Many of Which We Know

From German we know: Haus (house), Schule (school), Übermensch (superman), Wunderkind, Kuchen, Bratwurst, Kindergarten, Angst, Doppelgänger, Weltschmerz (world weariness), Weltanschauung (world view), Ansatz (educated guess that is later verified by its results), Ersatz (substitute), Gestalt (form or shape), Schadenfreude (pleasure derived from the misfortunes of others), Zeitgeist, Sturm und Drang ("Storm and Stress," a literary movement of the late eighteenth century), Blitzkrieg, Gesundheit, Donner und Blitzen (thunder and lightning), Dummkopf. ("Das ist nicht mein Haus, Dummkopf" hardly needs a translator.)

However, the German word *Lust* does not mean sexual desire. It means preference, as in, "Do you feel like going to the movies tonight?" In German the word *genial* means "of genius." But in English it means "friendly." In German the word *Roman* means "a novel." The German word *Sauerstoff* (sour stuff) means oxygen, but it derives from an eighteenth-century misconception of what makes acids acidic. They used to think that it was oxygen that made a chemical an acid. We now know that it is hydrogen that does.

In Mark Twain's lovely essay "The awful German language" he uses actual examples of German compound words and then hits us over the head with Freundschaftsbezeigungenstadtverordnetenversammlungenfamilieneigentümlichkeiten, which thankfully is actually jibberish. Incidentally, the longest word in literature comes from Aristophanes. The first sixty-five of the 179 letters needed for the English transliteration from Greek are: "Lopadotemachoselacogaleokranioleipsanodrimhypotrimmatosilphiopara...."[4]

German sentences can have multiple verbs at the end. *Schachtelsätze* ("box sentences") can look very strange. For example, "Die, die die, die die Anlagen beschädigen anklagen bekommen 50 euros." This means, "Those people who report the people who damage the parks will receive 50 euros."

An old joke goes something like this. Two German speakers are talking. "So, did you enjoy reading *Der Zauberberg* [The Magic Mountain] by Thomas Mann?" "I can't really say. My version came in two volumes. I've only read volume one, and all the verbs are in volume two."

A word pronounced "thonic" is perhaps the hardest word to spell in English: *chthonic*. It means "underground." A related word is *autochthonous*, which means "indigenous, native." But these words would not look so strange if your first language were Greek. They both include the combination of consonants $\chi\theta$.

The English word *serendipity* means finding something interesting or valuable by looking in the right place at the right time. It derives from the Persian word for the island Sri Lanka. It may be one of the top ten most difficult words to translate. Julius Comroe said, "Serendipity is looking in a haystack for a needle and discovering a farmer's daughter."[5]

Think of how many words in Japanese are familiar to us: sushi, tempura, wasabi, sake, kampai, banzai, kamikaze, harakiri, haiku, tsunami, sensei (teacher), shogun, samurai, Sudoku, Zen. *Subaru* is not just the car company; this is the Japanese name for the Pleiades star cluster. *Hashi* are chopsticks, but the primary meaning of this word is "bridge." If you lay two chopsticks across the top of your friend's glass, the message might be, "You drink alcohol like water flows under a bridge."

In Spain there is actually an institute that determines the officially correct way to say something in Spanish. This is the Real Academia Española. Related to this is the Asociación de Academias de la Lengua Española, which has chapters in many countries of Central and South America, the Philippines, and North America (Mexico and the United States).

In Spanish the word for wife is *esposa* and the word for wives is *esposas*. But *esposas* also means "handcuffs"!

The word "horde," as in "horde of barbarians," derives from *ordu*—a Mongolian word meaning "camp."

"Jerky" derives from *cha'arki*, which is from Quechua, the language of the Incas still spoken today in the Andes of South America.

Many artifacts of American black lingo derive not from Africa but from white indentured servants who came from England and worked with slaves in the American South before the Civil War. Examples: *aks* for ask; "she my best friend."[6]

The word "hysteria" derives from the Greek word for uterus.

In the world of Star Trek, we got to know Lieutenant Uhura. *Uhuru* (with U at the end) is the Swahili word for "freedom." It is also the name given to an X-ray satellite launched from Kenya in 1970.

The word "assassin" comes from thirteenth-century Azerbaijan, a country on the west side of the Caspian Sea. The rulers promised hired killers an afterlife of delights should they die on a mission. They received a taste of those delights prior to missions. These *hashishiyun* consumed a lot of hashish.[7]

The Spanish *ojala* (God willing) derives from Arabic (*inshallah*).

The Arabic word *hammam* means bathroom, but *hamaam* (with a shorter middle M sound and a longer A sound in the second syllable) means "pigeons." It is hard to hear the difference.

The Italian word *arrabbiata* (or with -o at the end for masculine singular) is an adjective meaning "angry." At a restaurant if you order *puttanesca con salsa arrabbiata* ("whore's pasta with angry sauce"), it will be very spicy.

Some words are just fun to say out loud, such as the French and Italian words for "cheese" (*fromage* and *formaggio*). There is the Russian word *sluzhba*, which means "obligatory military service" and the Yiddish word *shpilkes*, which means "a sense of anxiety and dread." Also, the Lithuanian word *rupūže*, where the last consonant is pronounced like "zh." This means "toad," but don't call anyone this unless you're really, really mad at them, because it's the worst thing you can call someone in Lithuanian.

In Russian, czar (or, more properly tsar) derives from Caesar. So, in a way the political tradition of ancient Rome lived on until 1918 when Czar Nicholas II was deposed.

The word "robot" originated in a play called *R. U. R.* or *Rossum's Universal Robots*, written in 1920 by the Czech writer Karel Čapek. Compare it to the Russian word *rabota*, meaning "work."

The essential slang word in Chile is *huevon*, which is usually said with the V silent. It derives from the word for eggs (*huevos*), slang for testicles. A guy who cuts you off in traffic gets yelled at and labeled a *maldito huevon*. But when your son wins the golf tournament and you give him a hug you might say, "*Oye, huevon.*" So the meaning of this word depends very much on the circumstances and one's tone of voice. The meaning ranges from "pal" to "you stupid, *stupid* jerk." One can make other related word forms. A totally messy ridiculous situation is a *hueveo* or *huevada* (in which the V is pronounced). *Corta el hueveo* means "cut the crap." You can make a reflexive verb (*hueviarse*). *Ayer mi jefe estaba hueviandome* means, "Yesterday my boss was messing with me."

In Hawaii they do not use directions north, south, east, and west very much. Rather, there is *mauka* (uphill, toward the mountain) and *makai* (downhill, toward the ocean). The other pair of directions is effectively clockwise around the island or counterclockwise. For that you need to know the names of some local places. On the east side of the Big Island, for example, Hilo is the most populous place. From there Honoka'a is north (counterclockwise) and Keaau is south (clockwise). Directions from Hilo might be "Go Honoka'a twelve miles" or, "Go Keaau six miles." But if you end up in Mountain View, on the other side of Keaau, then the direction "Keaau" is counterclockwise around the island, not clockwise.

In Hawaii the essential slang word is *dakine*, meaning "the person or thing we were talking about recently." You might say, "Well, go talk to dakine guy." And it is (probably) understood by both people who you are referring to. Of course, this can lead to some confusion, especially when it gets doubly nested. Someone might say to you, "Did you talk to dakine about dakine?" Did you talk to you-know-who about you-know-what? It begins to sound like two mob guys trying to be discrete on a phone that might be tapped. "Did you talk to the guy about the thing?"

There are toasts in various languages. We know *salud* or *salute* from Spanish and Italian, *Prost* from German. But how about this Russian toast: *dai Bog ne posledniuniu*. This means multiple things. 1) Dear God, may this not be the last time we get together. 2) May this not be the last drink. 3) Dear God, don't make me the guy in last place (the bottom guy on the totem pole). 4) Dear God, don't give me the last place things. In other words, give me *something* to be thankful for, please.

A good word to know in Zulu is *jabula* (rejoice). One toast in Zulu is *Jabula umphimbo uya phuzo* which means "Rejoice, my throat, you are about to drink." Thus, you toast your throat, not your drinking buddies. Useful phrases in Zulu can be found in reference[8] below. The weirdest not-so-useful example is this: My hovercraft is full of eels = *Umkhumbi wami ugcwele ngenyoka zemanzini*.

The Hawaiian word *aloha* means "hello" or "good-bye" or "love," depending on the context.

Endnotes

1. McWhorter, John, *What Language Is: (and what it isn't and what it could be)*, New York: Gotham Books, 2011, pp. 14, 139, 145–6.
2. Rice, Edward, *Captain Sir Richard Francis Burton: The Secret Agent Who Made the Pilgrimage to Mecca, Discovered the Kama Sutra, and Brought the Arabian Nights to the West*. New York: Charles Scribner's Sons, 1990.

3. http://en.wikipedia.org/wiki/James_Nicoll (accessed March 28, 2013).
4. Theroux, Alexander, *Darconville's Cat*, New York: Henry Holt and Company, 1996, p. 76.
5. http://writersalmanac.publicradio.org/index.php?date=2013/01/28 (accessed January 29, 2013).
6. McWhorter, ref. 1, p. 113 ff.
7. Sarton, George, *Introduction to the History of Science*, New York: Robert E. Krieger, 1975, vol. I, pp. 752–3.
8. http://www.omniglot.com/language/phrases/zulu.php (accessed December 29, 2012).

5

The Good, the Bad, and the Obsessive

The Most Brilliant Ph.D. Thesis Ever Written in Astronomy

The first person to earn a Ph.D. in astronomy from Radcliffe College (now part of Harvard) was Cecilia Payne (later Payne-Gaposchkin) in 1925.[1] Her dissertation, *Stellar Atmospheres*,[2] was "undoubtedly the most brilliant Ph.D. thesis ever written in astronomy."[3,4] In it she showed the uniform composition of the atmospheres of stars. The *spectra* differ, however, because of the wide range of their photospheric temperatures. For example, it requires less energy to strip a single electron from atoms of calcium, iron, sodium, and magnesium, compared to hydrogen. This is why we see strong lines of calcium, iron, sodium, and magnesium in the spectrum of the Sun, whose photosphere has a temperature of about six thousand deg K. A hotter star like Vega or Sirius (at ten thousand deg K) can easily ionize hydrogen. These two stars have much stronger hydrogen absorption lines in their spectra.

Cecilia Payne's work showed that many common elements in the Earth have the same abundances in the Sun. Most significantly, she showed that a considerable majority of the atoms in a star are hydrogen. This is not the case for the composition of the Earth. Curiously, toward the end of her dissertation she writes, "The outstanding discrepancies between the astrophysical and terrestrial abundances are displayed for hydrogen and helium. The enormous abundance derived for these elements in the stellar atmosphere is almost certainly not real."[5] She included this disclaimer under pressure from an outside reviewer, Princeton astronomer Henry Norris Russell. A few years later he realized that she had been right in the first place.

Yale astronomer Dorrit Hoffleit (1907–2007) told me this story about Cecilia Payne-Gaposchkin and her husband, Russian-born astronomer Sergei Gaposchkin. The two met at a meeting in Göttingen in 1933. He was desperate to get out of Europe. Stalin's Russia and Hitler's Germany were equally unacceptable. He asked if she could possibly pull any strings on his behalf. She thought about it for a while and said, "I'll see what I can do. But there are two conditions. First, we must get married. Second, we must have children." And so it came to pass. They got married and had three children.

Cecilia Payne-Gaposchkin became the first woman in the history of Harvard University to receive a corporation appointment with tenure; she also became the first woman department chair in 1956.[6] I have nothing but admiration for her. She was accomplished yet very humble. In fact, she had this perspective about credit where credit might be due: "It doesn't really matter how you give the credit for a scientific discovery, for if one person doesn't discover it, someone else will."[4]

The Most Egotistical Book in the History of Astronomy

At the other end of the scale of modesty and accomplishment was Missouri-born astronomer Thomas Jefferson Jackson See (1866–1962), who obtained his Ph.D. in mathematics from the University of Berlin in 1892 and became one of the first two astronomers at the University of Chicago the year it opened its doors (1892).[7] The other astronomer at Chicago was George Ellery Hale (1868–1938), the premier observatory builder of the turn of the twentieth century.[8] See left the University of Chicago in 1896 and spent two years associated with Lowell Observatory in Flagstaff, Arizona, then worked at the US Naval Observatory in Washington, DC. In 1903 he was transferred to Mare Island, California, to run the navy's weather station there.

I assert that the most egotistical book in the history of astronomy is the "biography" of T. J. J. See and account of his "unparalleled discoveries" by W. L. Webb.[9] While William Larkin Webb was a real person, it is highly likely that See himself wrote most or all of this book.

Consider the hyperbole. See "first saw the light on the birthday of Copernicus (1473–1543), the founder of modern astronomy, Feb. 19, 1866. This date of birth might be an accident, but the believers in astrology will find in the career of Professor See and his revolutionary work in astronomy so much to remind them of his great predecessor, as to cause many to think that after all our destinies are shaped by the stars under which we are born."[10] "The little boy of three never dreamed in this happy childhood that some day he was to be the one astronomer who could enlighten the world regarding the origin of the Moon.... Even in childhood he was every inch a natural philosopher."[11] "Little did his associates then imagine that the little boy with methodical methods and brilliant memory was to become the greatest astronomer in the world, and one of the greatest of all time!"[12] As a graduate student at the University of Berlin it was foreseen that See would become even more famous than his professor, Hermann von Helmholtz.[13] As a graduate student See's "fame spread to every department of the great University, and in fact all over Germany, and even to Italy, France, Russia, and England."[14] See's establishment of the "New Science of Cosmogony (1910) ... marks one of the greatest epochs in the history of astronomy."[15] "Some readers may not realize that the discoveries of Professor See ... will even outlast the Republic itself, and still be the topic of contemplation for philosophers when many

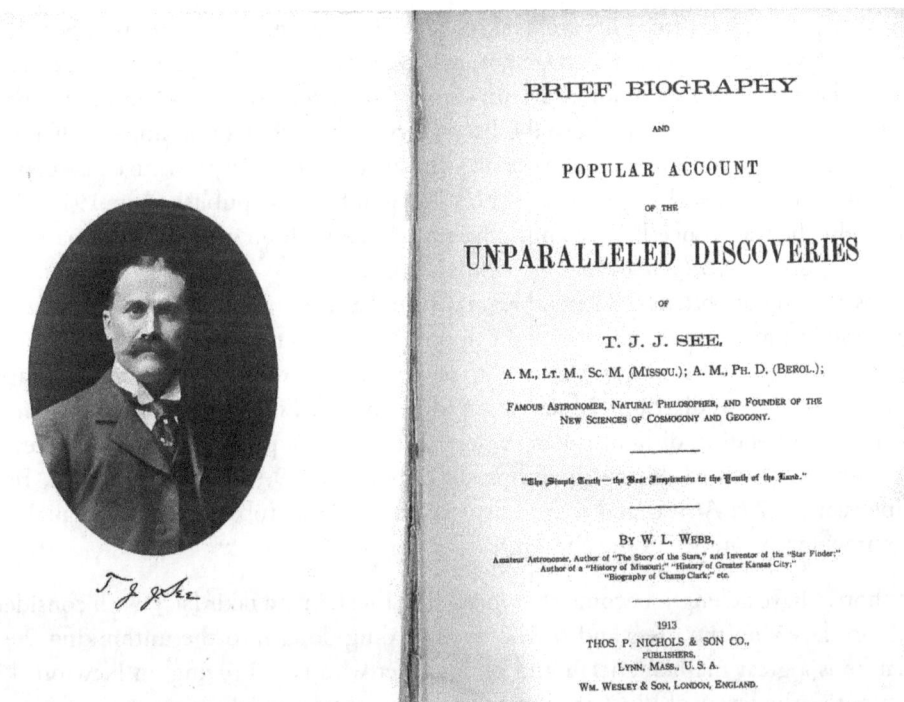

Figure 5.1 Frontispiece and title page of the "biography" of T. J. J. See. Note the quote, "The simple truth—the best inspiration to the youth of the land." The simple truth is that this book was probably not written by William Larkin Webb, a newspaper publisher, fellow Missourian, and admirer of See. This book is more likely the work of See himself.

thousands of years have elapsed; just as the works of Aristotle and Plato now belong not to Greece but to all mankind and to all time."[16]

If a biography or autobiography is to be believed, there has to be a downside somewhere, if the subject's life is known to have some controversy. But this is distinctly missing from the Webb "biography." Why did See leave the University of Chicago? Because he demanded a promotion similar to that given to George Ellery Hale. President William Rainey Harper decided that the University of Chicago could do fine without See. Why did See leave Lowell Observatory? Was he dismissed just for arrogance? Why was he banned from submitting articles to the *Astronomical Journal*? Any objective biographer would have adressed these issues.

In the spring of 1998 when the American Astronomical Society met in San Diego, California, I drove to Palomar Mountain for a ceremony honoring the fiftieth anniversary of the two-hundred-inch telescope. One of my fellow passengers was Dick Walker (1938–2005), who worked at the US Naval Observatory station in Flagstaff, Arizona. He told me that he had been looking at T. J. J. See material in the Lowell Observatory archives and had to promise not to show anyone his notes or photocopies from his archival research. He said he discovered that See had carried on some correspondence with a teenage boy, and this was part of the pretext for See's dismissal. Shades of the plot line from Thomas Mann's novella *Death in Venice*! When I asked permission to peruse the See material in the Lowell Observatory archives I was told no, and that *no one* will ever have permission to look at this material! Clearly, it must be very sensitive and controversial.

Unfortunately for See, by the time Webb's "biography" was published in 1913, "to almost everyone else, he was a once-licensed but now-suspended buffoon who had been barred from the pages of the *Astronomical Journal* for his vituperative excesses."[17]

See was a strong opponent of General Relativity and a strong proponent of the *aether*, which is the hypothetical medium required by light to propagate through the universe, disproven by the Michelson-Morley experiment of 1887.[18] See gave a number of public lectures on relativity and published a number of diatribes. Much of it had to do with Einstein's prediction of the amount of gravitational bending of light during a solar eclipse. This prediction had been tested using stellar positions obtained during the eclipse of May 29, 1919 (observed at Sobral, Brazil, and Principe Island, West Africa) and that of September 21, 1922 (observed in Australia). Consider this commentary from See in the *San Francisco Journal*:

> In short, I have at length become convinced that Einstein is a fakir [sic], with considerable skill in deceiving the press and public so as to ding-dong into the unthinking the idea that he is a great mathematician and philosopher who is improving on Newton. Let us first notice the errors of Einstein, and the cunning way in which he gets away from them, owing to the layman's inability to pin him down....
>
> In 1919 it was oracularly heralded abroad by Einstein that there is no aether. This pernicious proposition was echoed in Holland, and repeated in England by certain mediocre physicists in the Royal Society, more especially by Eddington and Jeans, who have since done so much to spread errors over the world.

See tried to get the associate director of Lick Observatory, Robert Aitken, to debate the issues in public or at least in the pages of San Francisco newspapers. Instead, Aitken returned See's papers to him, commenting that they were

> apparently expositions of your personal ideas and convictions, with some references to, and quotations from papers by other men.... Nor do I see that your letters require any special answer, though I might, perhaps, remind you that hear-say evidence is not accepted in a court of law and add that I question its weight with scientific men of standing. If you care to base public statements or arguments upon such evidence the responsibility must rest with you.

McDonald Observatory was funded by the will of a rich Texan and was built as a joint project of the Universities of Texas and Chicago. On this plan See bemoaned: "the Bungling and Unworkable plans you have formed with the Yerkes Observatory: it will not stand five years! It carries with it divided responsibility and divided credit for results, if any are attained, by the foreigner entrusted with matters which should be in the hands of a real American!"[19] See owned a farm in northeast Texas, which he was willing to sell to be the observatory site. See's opinion was that McDonald Observatory should have a large refractor and could be located at sea level. Instead, the Texas-Chicago agreement more sensibly led to the construction of the world's second largest reflector on a sixty-seven-hundred-foot peak in west Texas.

The personality of T. J. J. See reminds me of a quote attributed to the Russian writer Leo Tolstoy[20, 21]:

> Man is a fraction for which the numerator is what he actually amounts to, and the denominator is what he *thinks* he amounts to. The larger the denominator, the smaller the fraction, and if the denominator is equal to infinity, for whatever value of the numerator, the fraction is always equal to zero.

Though See had a huge desire to be a famous and accomplished astronomer, he had drummed himself out of the business in middle age. Part of his punishment was to live to the age of ninety-six, hearing not accolades from the world scientific community but silence.

Not for the Thin-Skinned

Next we turn to a bizarre episode related to the premier popularizer of astronomy toward the end of the nineteenth century, Camille Flammarion (1842–1925). I quote liberally from reference 22. In 1882 a young French countess and ardent fan of the author, dying of tuberculosis, wrote to her doctor:

> I have a confidential confession to make to you. I have loved Camille Flammarion with a flaming devotion, and now that I am dying I want him to have a souvenir from me. It will astonish you to know that I have never been presented to him, or talked to him, or even seen him, but I developed such an intense admiration for Monsieur Flammarion from reading his books and following his work that I secretly fell in love with him. I worshipped him day and night for five years. I want to remain with him, and so I beg you that as soon as I die, you cut a big piece of skin off my shoulders and send it to him as a binding for one of his books. I want my name kept a secret, however, and you must promise that if he comes and asks questions you will not reveal my identity.

Once the young woman died, the doctor, whose name was Ravaud, cut a twelve-by-eighteen-inch piece of skin from her body, rolled it up in a packet, and delivered it to Flammarion's house along with a note, which said:

> Monsieur:
>
> *True to my promise, I have carefully carried out the request of the dead Countess who always loved you. She begged me to send you, the day after her death, the skin of her lovely shoulders. This is the*

skin, and you must promise that you will use it to bind a copy of the first book you may publish now, after her death. I have delivered this souvenir to you, Monsieur, as I faithfully promised.

Dr. Ravaud

Flammarion went to see the doctor, who firmly stated that he would carry the identity of the woman to the grave with him, though he did admit that she was a member of one of the first families of France. And so it came to pass that Flammarion had the skin tanned, and it was used to bind a copy of his book *Terres du Ciel*. On the front cover it says the following (here translated from French): "Pious fulfillment of an anonymous wish, binding in human skin (woman) 1882."

Several years before Flammarion died, one Dr. Cabanès, editor of a French medical journal, received this account from the astronomer:

My Dear Doctor:

The story has been somewhat elaborated. I don't know the name of the person whose forsal skin was delivered to me by a physician to use for the binding. It was a matter of carrying out a pious vow. Some newspapers, especially in America, published the portrait, the name, and even the photograph of the chateau where "the Countess" dwelt. All of that is pure invention.

The binding was successfully executed by Engel, and from then on the skin was unchanging. I recollect I had to carry this relic to a tanner in the Rue de la Reine-Blanche, and three months were required for the job. Such an idea was assuredly bizarre. However, in point of fact, this vestige of a beautiful body is all that survives of it today, and it can endure lastingly in a perfect state of respectful preservation.

The desire of the unknown woman was to have my last book published at the time of her death bound in this skin: the octavo edition of the Terres du ciel, *published by Didier enjoys this honor.*

Your reader and admirer,

Flammarion.

The compiler of this account, Walter Hart Blumenthal, had some correspondence with the second Madame Flammarion concerning this matter. He received a photograph of the volume and evidence of its authenticity. More recently, my friend, Harvard professor Owen Gingerich, emeritus, has seen the book at the Flammarion estate at Juvisy-sur-Orage in northern France.

We still do not officially know the name of the young woman, but the Boston *Transcript* of May 31, 1924, identifies her as the Countess St. Agnès, while a contributor to *Notes and Queries* (1944, vol. 187) mentions the Countess de Saint-Auge.

Endnotes

1. Haramundanis, Katherine, editor, *Cecilia Payne-Gaposchkin: An Autobiography and Other Recollections*, Cambridge: Cambridge University Press, 2nd edition, 1996.

2. Payne, Cecilia H., *Stellar Atmospheres: A Contribution to the Observational Study of High Temperature in the Reversing Layers of Stars*, Cambridge, Masachusetts: Harvard University Press, 1925, Harvard Observatory Monographs, No. 1.
3. Struve, Otto & Velta Zebergs, *Astronomy of the 20th Century*, New York: Macmillan, 1962, p. 220.
4. Gingerich, Owen, "The most brilliant Ph.D. thesis ever written in astronomy," http://www.harvardsquarelibrary.org/unitarians/payne2.html (accessed February 1, 2013).
5. Payne, ref. 2, p. 188.
6. Miller, Heather, "Cecilia Payne-Gaposchkin: The bravery of a mind," http://www.harvardsquarelibrary.org/unitarians/payne2.html (accessed February 1, 2013).
7. Ashbrook, Joseph, "The sage of Mare Island," *Sky and Telescope*, 24, No. 4, October 1962, p. 193.
8. Wright, Helen, *Explorer of the Universe: A biography of George Ellery Hale*, New York: Dutton, 1966.
9. Webb, William L., *A Brief Biography and Account of the Unparalleled Discoveries of T. J. J. See*, Lynn, Massachusetts: Thomas P. Nichols & Sons, 1913.
10. Ibid., p. 10.
11. Ibid., p. 11.
12. Ibid., p. 13.
13. Ibid., p. 45.
14. Ibid., p. 48.
15. Ibid., p. 63.
16. Ibid., p. 87.
17. Evans, David A. & J. Derral Mulholland, *Big and Bright: A history of the McDonald Observatory*, Austin, Texas: University of Texas Press, 1986, p. 30.
18. Crelinsten, Jeffrey, *Einstein's Jury: The race to test relativity*, Princeton, New Jersey: Princeton University Press, 2006, pp. 216–225.
19. Evans & Mulholland, ref. 17, p. 31.
20. See *Tolstoy's Diaries: Volume I: 1847–1894* and *Tolstoy's Diaries: Volume II: 1895–1910*, edited and translated by R. F. Christian, New York: Charles Scribner's Sons, 1985.
21. Prof. Reginald Christian, the editor and translator of Tolstoy's diaries (ref. 20), corresponded with me about the "man is a fraction" quote. In a letter of November 27, 1992, he wrote: "I can't track down the original source of your quote, but I know occasions when he used the same metaphor in which the numerator stood for a person't positive qualities, and the denominator his own opinion of himself. He used it in his diaries about his son Seryozha ([April 21] 1894), his son Lyova ([November 10] 1897), and the writers Andreyev and Maxim Gorky (1909). I'm sure I've met it elsewhere too, but I don't know when Tolstoy first forumulated it." In Tolstoy's diary entry for February 4, 1909, he did not use the fraction metaphor but commented that the writings of Gorky and Andreyev were lacking in original thought, talent, and content, which would be the numerator part of the metaphor.
22. Blumenthal, Walter Hart, *An Olio Bookmen's Bedlam of Literary Oddities*, New Brunswick, New Jersey: Rutgers University Press, 1955, pp. 85–89.

6

How Long Do Astronomers Live?

How long do astronomers live? This is ostensibly a simple question, but it involves a particular conditional probability. If a baby does not die in childbirth and does not die of various childhood diseases, then once that baby has grown up and reached adulthood, he or she could live a long time. Once a person has become an astronomer, that string of conditionals might almost guarantee that the astronomer lives longer than the average person.

Furthermore, over time educated people who end up as scientists and professors might be expected to live longer than, say, oil rig workers, whose work is documented to be very dangerous.[1] Yet, some scientists work with hazardous radioactive substances, such as Marie Curie's work with radium. If you are a medical researcher and work with HIV or the Ebola virus, that would be considered a dangerous job.

There was an interesting *Time* magazine cover story that dealt with the causes of death in the modern-day United States.[2] The last year on record for that study was 2003. In that year 2.5 million people died in the United States: 685,089 died of heart disease; 556,902 died of cancer; 31,484 were suicides; 17,732 were murdered; 3,676 died in motorcycle accidents; and 44,757 died in other motor vehicle accidents. Commercial air travel is actually quite safe. In the previous ten years an average of eighty-two people per year died in commercial airline accidents. That included the 265 that died on September 11, 2001.

I have used my biographical index of the first 108 volumes of *Sky and Telescope* magazine as the database.[3] Using the version of the index as it existed in May of 2008, there were over seven hundred people in the database. Several years ago the magazine changed owners and one consequence was few, if any, obituaries.

A number of names should be eliminated from the statistics. Not everyone listed has a birth year and death year. Not everyone is an astronomer, such as William Shakespeare and Vincent van Gogh. Neither were Generals Thomas Brisbane or Milan Stefanik. But Geoffrey Chaucer should be retained because he wrote a treatise on the astrolabe. Finally, what should we do about the astronomers who were executed or died in prison? I think they should not be included in the statistics of how long astronomers live. Giordano Bruno (1548–1600) was burned at the stake by the Roman Inquisition. In the October 1989 issue of *Sky and Telescope* magazine Robert McCutcheon narrates the purge of Soviet astronomers in the late 1930s.[4] Some of those imprisoned outlived the Stalinist era, but at least nine astronomers in the database died in prison. Boris Petrovich Gerasimovich (1889–1937) was certainly executed.

As far as I know, eight (possibly nine) astronomers have been murdered in the normal sense of the word (i.e., not a victim of a Stalinist purge). The Neoplatonic philosopher, mathematician, and astronomer Hypatia (ca. 350–415 AD) was attacked by a riot-crazed mob of Christians who flayed the flesh from her bones using broken oyster shells then burned the pieces of her body.[5]

Ulugh Beg (1394–1449) was murdered by an assassin hired by his son.[6] This must have been related to the fact that he had become the local prince.

Auguste Charlois (1864–1910) was an asteroid hunter who discovered ninety-nine asteroids. Many were apparently named after his mistresses. He eventually got divorced from his long-suffering wife. But when he was about to remarry, his ex-brother-in-law killed him.[7] Incidentally, Charlois is not in the *Sky and Telescope* index.

Four members of the Department of Physics and Astronomy at the University of Iowa were murdered on November 1, 1991, by a former graduate student of that department. The four were Christoph Goertz, Dwight Nicholson, Robert A. Smith, and Linhua Shan. The perpetrator was Gang Lu, who was infuriated because he did not receive a prestigious prize for his dissertation. A film based on this set of events, *Dark Matter*, won a prize at the 2007 Sundance Film Festival.[8]

Australian astrophysicist Rodney Marks died suddenly at the South Pole on May 12, 2000. His body could only be flown to New Zealand for an autopsy several months later. The cause of death was methanol poisoning. It might have been murder.[9]

Japanese astronomer Koichiro Morita was killed on May 7, 2012, while walking down the street in Santiago, Chile. He was set upon by a robber who wanted his wallet.[10]

For the year 2003 in the United States the fraction of deaths by murder was 17,732 out of 2.5 million, or 0.7 percent. Charlois, Marks, Morita, and the victims of the University of Iowa shootings are not in the *Sky and Telescope* name index, so maybe we should should quote two out of seven hundred, or 0.3 percent as the murder rate of astronomers. Maybe we should consider the cause of death of a number of the Soviet astronomers who perished in the Stalinist purge of the 1930s to be murder.

In the general population the suicide rate is 78 percent higher than the murder rate, but amongst astronomers it may be significantly higher. This subject is too sad if I consider people I knew personally. Of notable astronomers and physicists of previous eras who died by their own hand we mention the founder of statistical mechanics, Ludwig Boltzmann (1844–1906),[11] Friedrich Wilhelm Ristenpart (1868–1913), one-time director of the Chilean National Observatory,[12] and William Wallace Campbell (1862–1938), a director of Lick Observatory.[13]

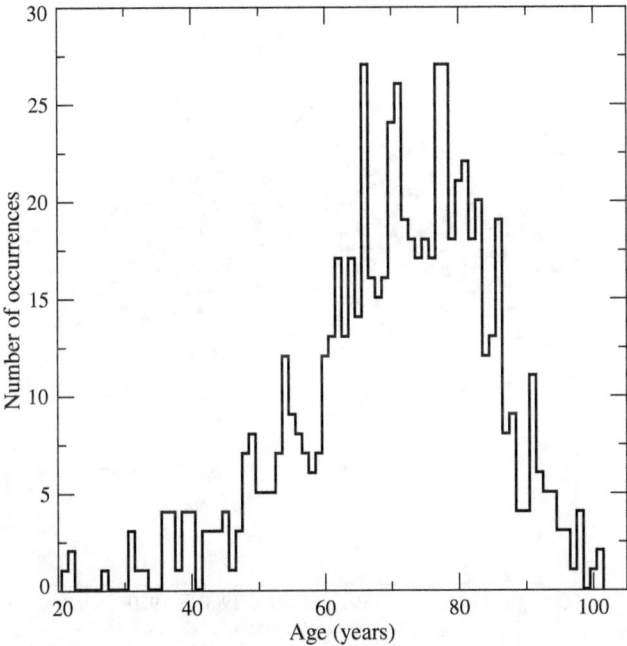

Figure 6.1 Histogram of life span of 691 astronomers from a biographical index to the first 108 volumes of *Sky and Telescope* magazine.

In Figure 6.1 we show a histogram of 691 astronomers and related people from the *Sky and Telescope* index. (Shakespeare, van Gogh, and two generals mentioned above were excluded.) The median life span was seventy-two. The distribution is clearly asymmetric; this only means that there are many ways to die considerably younger than the median, while at the high end of the distribution we are up against the maximum life span possible for humans.

Figure 6.2 shows the individual values of life span for astronomers born in the year 1800 and later. The nine squares correspond to astronomers who were executed or otherwise died in prison as a result of the Soviet purge of the late 1930s. The solid line at the right edge is significant. Say there was an astronomer born in 1920 who lived to be 85 and whose death was noted in *Sky and Telescope* magazine. That astronomer could not have been part of the database in 2004, the last year indexed. Given that astronomers have lived to more than a hundred, the database is more and more incomplete starting with birth year 1903.

There are three astronomers in the database who lived to one hundred or 101. They were Charles Greeley Abbot (1872–1973), Goethe Link (1879–1980, who was a physician and amateur astronomer who donated his thirty-six-inch telescope to Indiana University), and Giorgio Abetti (1882–1982). Three long-lived astronomers not in the database (as they were not noted in *Sky and Telescope* magazine) were Dorrit Hoffleit, who died in April 2007 at the age of one

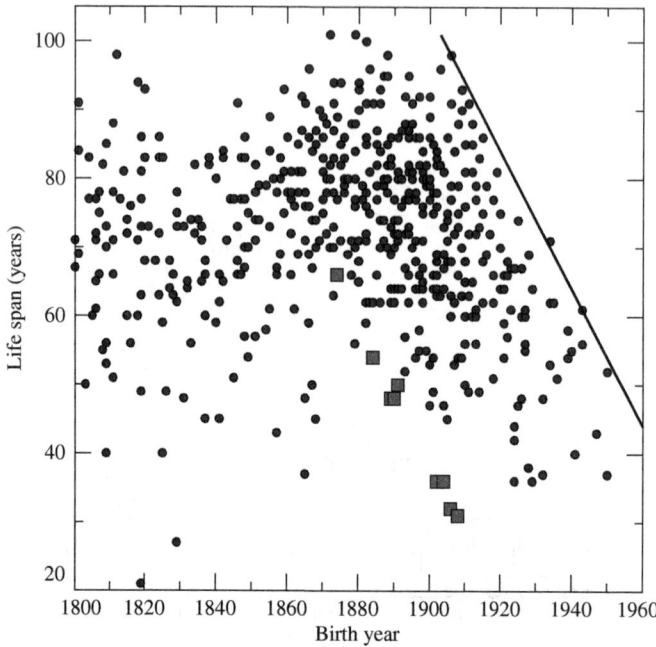

Figure 6.2 Age at death vs. birth year for astronomers born from 1800 through 1960. The squares correspond to nine Soviet astronomers who were imprisoned or executed during the Stalinist purge of the 1930s.

hundred and one month, Theodore Jacobsen (1901–2003), and until recently the longest-lived astronomer, Paul Sollenberger, who died in 1995 at the age of 103 and nine months.[14] Yoshio Fujita died on January 9, 2013, at the age of 104. To my knowledge he is now the longest-lived astronomer. An obituary will be written for the *Bulletin of the American Astronomical Society*.[15]

The longest-lived person associated with anything astronomical was Live Larsdatter, who worked for Tycho Brahe on Hven or in Copenhagen. She died in 1698, supposedly at the age of 122 years and eleven months.[16] Perhaps we should consider this "fact" with a grain of salt, but there *are* credible claims of people living to the age of 122.[17]

In Table 6.1 we give the average values of life span of the astronomers in the database. We bin according to the year of birth. For the fifteenth, sixteenth, and seventeenth centuries we give the full century averages. Then we use fifty-year bin widths. Finally, we use ten-year bin widths from the 1840s through the first two decades of the twentieth century. We also used a sliding window, incremented by one year at a time, to determine which ten-year period gives the absolute maximum. The last decade that is reasonably complete is 1901–1910.

In Figure 6.3 we show a subset of the entries from Table 6.1. The maximum life span of astronomers occurred for the ten-year bin 1869 to 1878, when the mean life expectancy was 80.63. This is plotted as a triangle. A line is fitted to the dots. The point for the 1901 to 1910

Table 6.1 Mean Life Span of Astronomers

Period	N	Mean[a]	Median
315BC–1400	14	71.64 (2.98)	73
1401–1500	6	61.83 (4.98)	65
1501–1600	21	64.29 (1.83)	65
1601–1700	35	67.40 (2.77)	70
1701–1750	39	68.69 (2.32)	71
1751–1800	52	66.71 (2.12)	69.5
1801–1850	114	69.44 (1.25)	71
1841–1850	26	70.92 (2.21)	70.5
1851–1860	27	74.74 (2.08)	78
1861–1870	43	75.44 (1.97)	77
1867–1876	46	79.41 (1.54)	80.5
1868–1877	44	80.16 (1.44)	80.5
1869–1878	41	80.63 (1.30)	80
1870–1879	48	80.15 (1.32)	80
1871–1880	50	80.04 (1.42)	80
1872–1881	48	80.00 (1.46)	80
1873–1882	55	79.36 (1.39)	79
1881–1890	64	77.94 (1.25)	78
1891–1900	73	74.67 (1.32)	77
1901–1910	63	71.35 (1.55)	71
1911–1920	34	70.65 (1.91)	72

[a] Numbers in parentheses are the "mean error of the mean" ($1 - \sigma$ random errors).

bin is entirely consistent with an extrapolation of the mean values from 1401 to 1850, but the astronomers born between 1851 and 1900 lived statistically longer than astronomers in any other time period before or since.

The reasons for the excess of long-lived astronomers from the second half of the nineteenth century are not entirely clear. A man born in the 1870s was generally too old to have

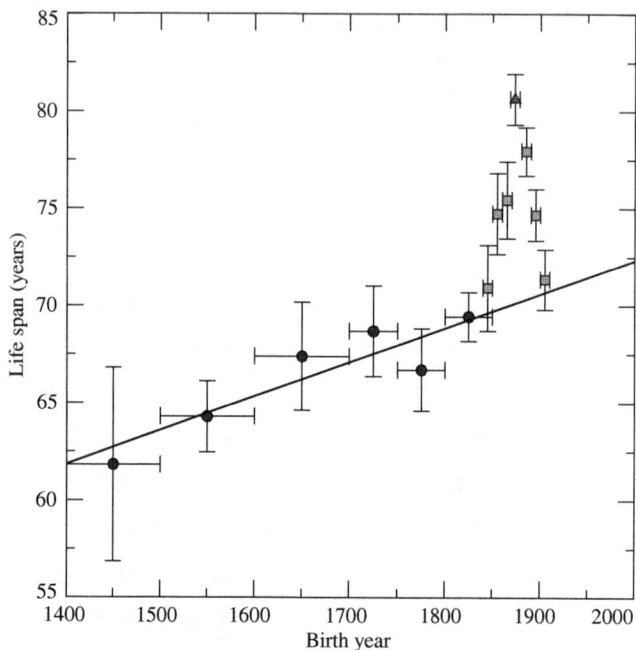

Figure 6.3 Binned values of life span. We show a subset of the entries from Table 6.1. Some bins are one hundred years wide, while others are fifty or ten years wide. The average for the ten-year period 1869 to 1878 (triangle) corresponds to the maximum mean life span of 80.6 years.

served in World War I. One exception was the physicist Karl Schwarzschild (1873–1916), who volunteered to be a solider when he was forty.

Some astronomers born at the very end of the nineteenth century did see military service. Ukrainian-born American astronomer Otto Struve (1897–1963) served in the Russian Imperial Army during World War I, then with the White Russians against the Bolsheviks during the Russian civil war of the early 1920s. The cause of his death was cirrhosis of the liver, which came about as a result of hepatitis he had contracted as a soldier in World War I.[18]

If an astronomer were born in the second half of the nineteenth century and avoided military service, what was the most dangerous thing he or she did? Travel to mountaintop observatories? No. With the exception of Lick Observatory (founded 1888), there were no mountaintop observatories. Going up and down the ladder in the campus observatory dome at night was probably the most dangerous activity.

It could be that our *Sky and Telescope* index is biased somehow, and the trend shown in Figure 6.3 is spurious. If we consult the obituaries of members of the American Astronomical Society, we note that they only begin in earnest in 1989. To make a histogram of the life spans by birth year truncates the *short* end of the histogram for astronomers born in the early decades of the twentieth century. We will not easily be able to say what was the mean life span of astronomers born in the first half of the twentieth century until the year 2060 or so.

Endnotes

1. Wang, Yang, & Olsen, Lise, "Deadliest job in Texas," *Houston Chronicle*, February 11, 2013, pp. A1, A18–A19.
2. Kluger, Jeffrey, "Why we worry about the things we shouldn't ... and ignore the things we should," *Time*, December 4, 2006, pp. 64–71.
3. http://people.physics.tamu.edu/krisciunas/st.html (accessed November 13, 2013).
4. McCutcheon, Robert, "Stalin's purge of Soviet astronomers," *Sky and Telescope*, 78, no. 4 (October 1989), p. 352.
5. Krisciunas, Kevin, *Astronomical Centers of the World*, Cambridge: Cambridge University Press, 1988, p. 19.
6. Ibid., pp. 32–34.
7. Simonin, M., "Necrologie: Auguste Charlois," *Astronomische Nachrichten* 184, 191–192 (1910). A list of asteroids discovered by Charlois is at http://en.wikipedia.org/wiki/Auguste_Charlois (accessed February 11, 2013). How many of these are named after mistresses is uncertain, but there certainly are a lot of female names.
8. http://en.wikipedia.org/wiki/University_of_Iowa_shooting (accessed April 15, 2013).
9. Panek, Richard, *The 4% Universe*, Boston, New York: Houghton Mifflin Harcourt, 2011, pp. 211–212.
10. http://www.bbc.co.uk/news/world-latin-america-18043856 (accessed February 11, 2013).
11. *Dictionary of Scientific Biography*, 2, pp. 260–268.
12. Duerbeck, Hilmar W., "National and international astronomical activities in Chile (1849–2001), *Acta Historica Astronomiae*, 14, pp. 204–234 (2002).
13. Osterbrock, Donald E., John R. Gustafson & W. J. Shiloh Unruh, *Eye on the Sky: Lick Observatory's First Century*, Berkeley, California: University of California Press, 1988, p. 148.
14. Dick, Steven J., *Bulletin of the American Astronomical Society*, 27, no. 4, 1995, pp. 1482–3. Also: http://aas.org/obituaries/paul-sollenberger-1891-1995 (accessed February 11, 2013).
15. http://aas.org/obituaries/yoshio-fujita-1908-2013 (accessed February 11, 2013).
16. Christianson, John Robert, *On Tycho's Island: Tycho Brahe and His Assistants, 1570–1601*, Cambridge: Cambridge University Press, 2000, pp. 311–312.
17. http://en.wikipedia.org/wiki/List_of_the_verified_oldest_people (accessed February 14, 2013). See also http://www.dailymail.co.uk/news/article-2198608/ (accessed February 13, 2013).
18. Krisciunas, Kevin, "Otto Struve: August 12, 1897–April 6, 1963," *Biographical Memoirs of the National Academy of Sciences*, 61, pp. 350–387 (1992).

7

Astronomical Measurements without a Telescope

In this chapter we describe some astronomical experiments that can be carried out with very simple equipment. Details of our own work can be found in two articles published in the *American Journal of Physics*.[1,2]

Aristarchus (ca. 287–212 BC) devised a clever method for determining the distance to the Moon using the geometry of a lunar eclipse. Such an eclipse happens when the full Moon occasionally enters the Earth's shadow. This does not happen every month because the plane of the Moon's orbit around the Earth is inclined by five degrees to the plane of the Earth's orbit around the Sun. Aristarchus determined that the distance to the Moon is between sixty and seventy Earth radii. One fact necessary for the method of Aristarchus is the mean angular size of the Moon. This is about 0.5 degree.

Claudius Ptolemy (ca. 100–170 AD) summarized Greek astronomy in his famous book, the *Almagest*. In it he states that the angular size of the Moon ranges from 31.3 to 35.3 arc minutes, where sixty arc minutes equals one degree, by definition. However, Ptolemy was primarily interested in describing the *direction* toward the Moon. His model of the motion of the Moon implied that it ranged in distance from 33.55 Earth radii to 61.17 Earth radii, nearly a factor of two in distance. Thus, its angular size should also range by a factor of two, rather than ±6 percent. And it was Ptolemy's model of the *motion* of the Moon that was used until it was supplanted in the sixteenth century.

I got to thinking, "Using the kind of equipment that Ptolemy had at his disposal [i.e., no telescopes], is it possible to measure the cyclical variations of the Moon's angular size?" Very few such measurements have come down to us from antiquity or the Middle Ages. Two observers who addressed this issue were Levi ben Gerson (1288–1344) and Ibn al-Shatir (1304–1375/6). Like Ptolemy, they claimed that the Moon's angular size varies by a few arc minutes.

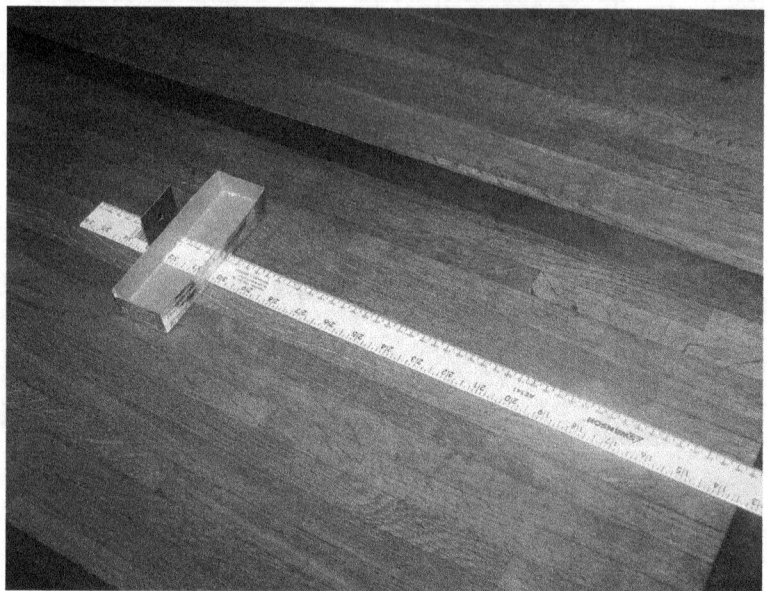

Figure 7.1 Device for measuring the angular size of the Moon.

I fashioned a Moon-sighting device from the bottom of a box of pancake mix, a second piece of cardboard, and a metal yardstick (see Figure 7.1). After observations over seven cycles of the Moon's phases, I had enough data to demonstrate that regular variations of the Moon's angular size *can* be measured with the naked eye.[1] Since then I have taken more data. Over a span of 1145 days (nearly thirty-nine cycles of the Moon's phases) I obtained a hundred measurements.[3] Our mean value of the Moon's angular diameter was 31.1 arc minutes, essentially equal to the accepted modern value. The data yield a mean perigee to perigee period of 27.5042 ± 0.0334 days. The modern, officially correct, value is 27.55455 days. (When a body that orbits the Earth is as close as it gets, it is said to be at perigee.) The agreement is quite satisfactory.

In Figure 7.2 we show our data phased with the derived period. Clearly, there is considerable scatter. Any individual data point is uncertain, on average, by ± 0.9 to 1.0 arc minutes, but we were able to demonstrate how much the Moon's angular size (and hence distance) varies. We found a value of the eccentricity of the Moon's orbit of 0.039 ± 0.004, which is a bit smaller than the official modern value of 0.0549. Due to the extra effect of the Sun's gravitational force on the Moon, the Moon's distance actually varies from +5.8 percent to −7.3 percent from the mean value.[4]

Another device from olden times is the *gnomon*, like the one pictured in Figure 7.3. It is basically just a vertical stick that casts a shadow of the Sun on the ground. To determine the

7 Astronomical Measurements without a Telescope

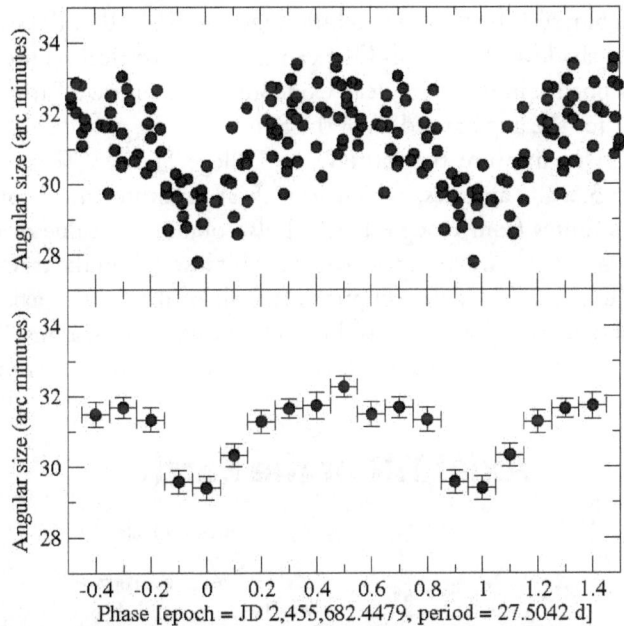

Figure 7.2 Phased observations of the angular size of the Moon. The upper diagram shows the individual data points obtained from April 21, 2009, through June 9, 2012. The lower diagram shows the averages for binned data.

Figure 7.3 Device for measuring the elevation angle of the Sun above the horizon.

elevation angle of the Sun, it is better to use a little sphere at the end of the stick. One marks the center of the elliptical shadow of the ball. One can use just a vertical pointed stick or a pointy statue, like the Luxor obelisk in the Place de la Concorde in downtown Paris, but then one must remember to account for the angular radius of the Sun.

I used this gnomon to measure the latitude of College Station, Texas, obtaining a value of 30 degrees 48.7 ± 3.5 arc minutes, which is 11.5 arc minutes north of the true value of 30 degrees 37.2 arc minutes from Google Earth. I also obtained a value of the *obliquity of the ecliptic* of 23 degrees 04.7 ± 3.5 arc minutes, which is a bit lower than the true value of 23 degrees 26.2 arc minutes. The obliquity of the ecliptic is the tilt of the Earth's axis of rotation to the plane of its orbit, as shown in Figure 7.4. It is this tilt that causes our seasons. When the northern hemisphere is tilted toward the Sun by 23.4 degrees, it is the beginning of summer in the north.

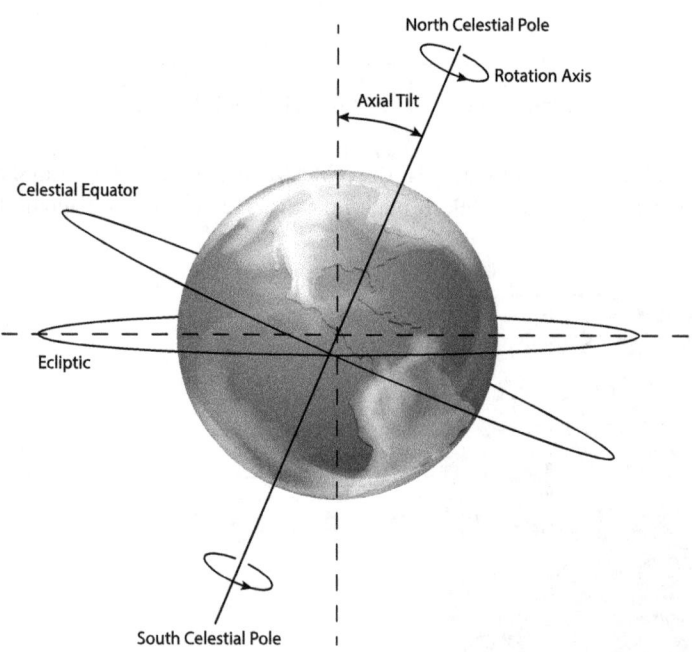

Figure 7.4 The Earth in space. The horizontal plane represents the plane of the Earth's orbit around the Sun. The axis of rotation of the Earth is not perpendicular to that plane. It is tilted 23.4 degrees, as shown. The axis of rotation points toward the South Celestial Pole (SCP) in the southern hemisphere, and toward the North Celestial Pole (NCP) in the northern hemisphere. *Image © BlueRingMedia, 2013. Used under license from Shutterstock, Inc.*

When the northern hemisphere is tilted away from the Sun by 23.4 degrees, it is the beginning of winter in the north.

Having determined my latitude and longitude in South Bend, Indiana, and College Station, Texas, I was able to determine a value of the circumference of the Earth, some 24,557 miles. The resulting radius of the Earth is 6290 km, which is 1.4 percent less than the official equatorial radius of the Earth of 6378 km.

Lahaye (2012) has elaborated a method of determining the eccentricity of the Earth's orbit using a gnomon.[5] This involves the *time* at which the Sun is highest in the sky, thus producing the minimum gnomon shadow length. This is not the same time on your watch, day after day. The *mean* solar time ranges from fourteen minutes ahead of *apparent* solar time to sixteen minutes behind it.[6] This is a result of the tilt of the Earth's axis of rotation and the eccentricity of the Earth's orbit. A full proof of this involves trigonometry and considerations of the geometry of an ellipse, which we need not repeat here.

Lahaye obtained measurements of the maximum elevation angle of the Sun at his location and the time of apparent noon. He did this four or five times each month throughout the year. He obtained a value of the obliquity of the ecliptic of 23.5 ± 0.1 degrees. For the eccentricity of the Earth's orbit he obtained 0.017 ± 0.001; the modern accepted value is 0.0167. From six gnomon experiments carried out on the campus of Texas A&M University and our value of the obliquity of the ecliptic, we obtained 0.014 ± 0.003 for the eccentricity of the Earth's orbit.

Throughout history astronomers have been preoccupied with the distance scale of the universe. This involves calibration of many rungs on what is called the *cosmological distance ladder*. This begins with surveying planet Earth—determining how big it is. We can then move on to determine the distance to the Moon in terms of the radius of the Earth. Using simultaneous observations of Mars or an asteroid from two locations on the Earth, we can then calibrate the scale of the solar system. This third rung of the distance ladder requires the use of telescopes. Prior to the end of the seventeenth century we only knew the *relative* sizes of the orbits of the planets, according to Kepler's Third Law of planetary motion. We discuss the first three rungs of the distance ladder in our 2012 article.[2]

To determine the distances to the nearest stars other than the Sun requires positional measurements considerably more accurate than one second of arc (1/3600 of a degree). We use the diameter of the Earth's orbit as our baseline for surveying the cosmos. This method of determining distance is called *trigonometric parallax*. The nearby stars move back and forth with respect to the distant background of stars owing to the motion of the Earth around the Sun. The first parallax measures were not obtained until the 1830s, nearly three centuries after Copernicus suggested that the Earth was just another planet orbiting the Sun.

The calibration of the scale of our Galaxy and the rest of the universe is carried out with *astronomical standard candles*, objects whose intrinsic brightness we know by one method or another. Two key types are pulsating stars called Cepheids, whose periods of pulsation are related to the mean brightness, and Type Ia supernovae, which are exploding white dwarf stars visible halfway across the universe with 4-m class telescopes. At maximum brightness such a supernova is roughly four billion times more luminous than the Sun.[7]

Endnotes

1. Krisciunas, Kevin "Determining the eccentricity of the Moon's orbit without a telescope," *American J. of Physics*, 78, pp. 828–833 (August 2010).
2. Krisciunas, K., E. DeBenedictis, J. Steeger, A. Bischoff-Kim, G. Tabak & K. Pasricha, "The first three rungs of the cosmological distance ladder," *American J. of Physics*, 80, no. 5, pp. 429–438 (May 2012).
3. The individual data points and further discussion can be obtained at http://people.physics.tamu.edu/krisciunas/moon_ang.html (accessed November 13, 2013).
4. *Allen's Astrophysical Quantities*, 4th ed., Arthur N. Cox, ed., New York, Berlin, Heidelberg: Springer-Verlag, 2000, p. 308.
5. Lahaye, Thierry, "Measuring the eccentricity of the Earth's orbit with a nail and a piece of plywood," http://arxiv.org/abs/1207.0982 (2012).
6. http://en.wikipedia.org/wiki/Equation_of_time (accessed March 9, 2013).
7. Krisciunas, Kevin "The usefulness of Type Ia supernovae for cosmology—a personal review," *J. of the Amer. Assoc. of Variable Star Observers*, 40, pp. 334–347 (May 2012). Also available via: http://arxiv.org/abs/1205.6835

8

Improvements in Astronomical Imaging

One of the first discoveries made by Galileo in the winter of 1609–10 was that a telescope reveals stars to us that are fainter than the eye can see without the telescope. The band of the Milky Way is mostly made up of faint stars. He described this in one of the first scientific best sellers, his book *The Sidereal Messenger*. He wrote, "For the Galaxy is nothing else than a congeries of innumerable stars distributed in clusters."[1]

While it is the *focal length* of a telescope that provides the magnification, there are practical limits to magnification that can be achieved. For a fifteen-centimeter (six-inch) diameter of ninety-centimeter focal length, one can rarely use eyepieces that give more than 150 power. Having a longer focal length telescope of the same diameter does not really help you much.

The German-born English astronomer William Herschel (1738–1822) was one of the first people to grasp the importance of the *collecting area* of the telescope and how that is effectively the *light-gathering power*. He strove to build reflecting telescopes with the largest possible diameter. His most productive telescope was one of focal length twenty feet, with a diameter of 47.5 centimeters (18.7 inches).[2,3] If we have two telescopes of similar design, and one is twice the diameter of the other, the larger telescope allows you to see stars that are four times fainter, because its objective has four times the area of the objective of the smaller telescope.

Until the beginning of the twentieth century, most astronomers worked with refracting telescopes (the kind with the lens up front). They were mostly interested in measuring the positions of the stars with ever greater precision. Herschel, on the other hand, was more interested in mapping the whole sky, discovering double stars, nebulae, and star clusters. He speculated on the structure and evolution of the Galaxy. For this he needed to be able to detect the faintest stars possible. So he cast mirrors out of the kind of metal used to build church bells. Finally, in 1857 it was possible to make a glass telescope mirror whose front surface was coated in a layer of silver.[4]

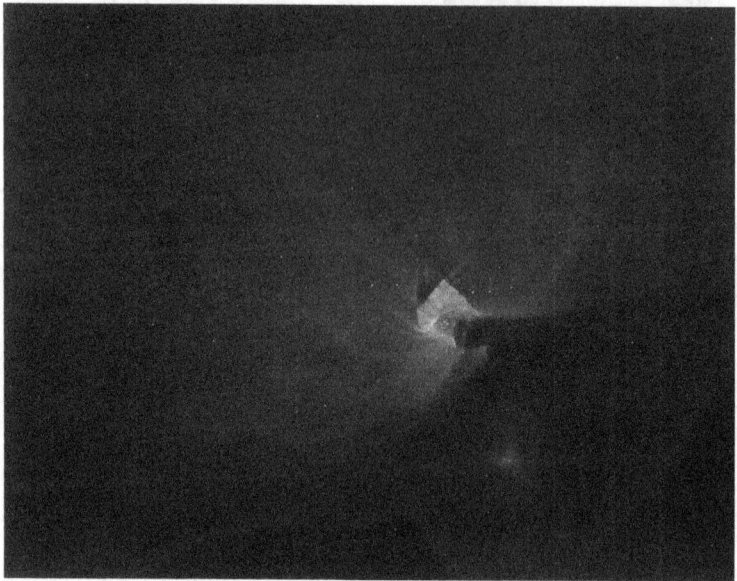

Figure 8.1 Drawing of the Orion Nebula by George Phillips Bond, 1859–1863.

Consider the drawing of the Orion Nebula shown in Figure 8.1. This was made by the Harvard astronomer George Phillips Bond (1825–1865) using the fifteen-inch diameter refractor of Harvard College Observatory and his eye.[5]

Photography was invented about 1838 by Louis Daguerre and Joseph Niepce.[6] At first only the Sun or Moon could be imaged, but by 1882, photography had improved to the point that images of the Orion Nebula could be made.[7] In Figure 8.2 we see a 137-minute exposure by Henry Draper. It revealed stars to magnitude 14.7, which is three thousand times fainter than the naked eye can detect.

One of the first successful professional reflecting telescopes was the thirty-six-inch diameter Crossley telescope.[8] It was acquired from a rich English amateur and reengineered to work at the latitude of Lick Observatory. With it the second Lick director, James Keeler, photographed many faint nebulae and clusters. And in 1908 a sixty-inch diameter reflector came on line at Mt. Wilson Observatory near Pasadena, California. By this point reflectors had supplanted refractors as the telescopes of choice for professional astronomers.[9]

Photography continued to improve up to 1970. But the *quantum efficiency* (or QE) of photographic emulsions has never been very impressive. The QE is the efficiency with which light is converted into an image on film, or the efficiency of converting light into electrons, which can be amplified in a photomultiplier tube or accumulated in the silicon layer of a solid state detector. Typically, photographic plates never exceeded a QE of 3 percent, though that could be increased

Figure 8.2 Photograph of the Orion Nebula, by Henry Draper, 1882.

Figure 8.3 Photograph of the spiral galaxy Messier 81, by James Keeler, using the thirty-six-inch Crossley reflector at Lick Observatory.

to 10 percent with hypersensitization.[10] Today, astronomers use charge coupled devices (CCDs), which can achieve QEs greater than 80 percent at some wavelengths. This allows a forty-inch diameter telescope to take images deeper than one could take in the 1960s with the Palomar two-hundred-inch telescope using photographic plates.

Figure 8.4 Supernova 2012fr is seen just above the central bulge of NGC 1365 in this combination ultraviolet/optical image from the Swift satellite. Courtesy Peter Brown, Texas A&M University.

Endnotes

1. *Sidereus Nuncius, or The Sidereal Messenger: Galileo Galilei,* translated by Albert van Helden, Chicago and London: University of Chicago Press, 1989, p. 62.
2. Riekher, Rolf, *Fernrohre und ihre Meister,* Berlin: Verlag Technik, 2nd ed., 1990, p. 138, and private communication from Dieter Herrmann to Kevin Krisciunas, March 11, 2013.
3. Herrmann, Dieter B., *The History of Astronomy from Herschel to Hertzsprung,* translated by Kevin Krisciunas, Cambridge: Cambridge University Press, 1984, pp. 7–14, 159–160.
4. Ibid., p. 167.
5. Holden, Edward S., *Monograph of the Central Parts of the Nebula of Orion,* Washington Astronomical Observations for 1878—Appendix I, Washington, DC: Government Printing Office, 1882. The Bond drawing is the frontispiece.
6. Herrmann, ref. 3, pp. 81–85.
7. Holden, ref. 4. The Draper photograph follows p. 226.

8. Keeler, James Edward, *Photographs of Nebulae and Clusters made with the Crossley Reflector*, Publications of the Lick Observatory, vol. VIII, 1098. The photograph of M81 is plate 21.
9. Osterbrock, Donald E., "The quest for more photons: how reflectors supplanted refractors as the monster telescopes of the future at the end of the last century," *Astron. Quarterly*, 5, pp. 87–95 (1985).
10. http://www.scribd.com/doc/6715280/Photographic-Plates-vs-CCD-in-Astronomy (accessed February 23, 2013).

9

The Constellations

The star catalogue presented in Ptolemy's *Almagest* (second century AD) was probably based on the earlier catalogue of Hipparchus (second century BC). There were forty-eight constellations. Some date back many thousands of years.[1,2] The far southern constellations were always below the horizon to observers in the Mediterranean so were absent from Ptolemy's catalogue.

Over the centuries some constellations (like Argo Navis) have been broken up and renamed. Most of the southern constellations were named in the eighteenth century by Nicolas Louis de Lacaille (1713–1762). At that time the Industrial Revolution was in full swing, and many of the new constellations were named after pieces of machinery, such as Antlia (the air pump), Microscopium, and Telescopium.

Ian Ridpath has written an excellent account of the history of constellations.[3] This includes obsolete constellations, such as the unfortunate sounding Turdus Solitarius. This constellation resembles a female blue rock thrush (*Monticola solitarius*, family *Turdidae*).

The modern constellation boundaries were organized by Belgian astronomer Eugène Delporte (1882–1955). As Ridpath describes:

> Delporte's new boundaries were approved by the [International Astronomical Union] at its General Assembly at Leiden in 1928. The Assembly also asked him to modify Gould's southern hemisphere boundaries to make them consistent with the new scheme for the north. This he did, in particular removing the diagonal lines that Gould had occasionally used and replacing them with verticals and horizontals. The final job was published in 1930 in *Délimitation Scientifique des Constellations* and an accompanying volume, *Atlas Céleste*.[4]

A translation of the introduction to *Dèlimitation* can be found in reference 5. Perhaps it is just a coincidence, but the number of constellations is the same as the number of keys in a regular piano keyboard.

The Pirates of Penzance (1879), by Gilbert and Sullivan, contains a tongue twister called The Major General's Song. Nearly a century later, mathematician and clever rhymster Tom Lehrer rewrote it to list the periodic table of the elements.[6] With that as piano music you may play my 1987 rewrite, which has the eighty-eight constellations in it. Two do not scan at all: Canes Venatici (the Hunting Dogs) and Camelopardus (the Giraffe). And you have to get ahead of the beat with Sagittarius. So far as I know, I am the only person who has ever performed this song in public.

The constellations in the sky were not put there to bore us.
There's Antlia, Andromeda, Aquarius, and Taurus,
And Apus, Leo, Lepus, Ara, Aries, and Centaurus,
Carina, Crater, Cancer, Canis Major/Minor, Corvus,
And Aquila, Auriga, Leo Minor, and Monoceros,
Columba, Libra, Lyra, Lynx, and Caelum, Cygnus, Circinus,
Corona Borealis, Cetus, Crux, Oh! do not scorn us
When you learn we fondly love the likes of Capricornus.

Horologium and Hydra, Hydrus, Indus, Microscopium,
And Cepheus, Bootes, Eridanus, Telescopium,
Chamaelion and Hercules, Dorado, Draco, Delphinus,
And Norma, Octans, Lupus, Musca, Pyxis, Grus, and Scorpius.

There's Coma Berenices, Fornax, Pavo, Puppis, Perseus,
Corona Australis, Tucana, places like Equuleus,
Sagittarius and Sagitta, both north and south Triangulum,
And Scutum, Sculptor, Sextens, Serpens, Mensa, and Reticulum.
Lacerta, Ophiuchus, Pictor, Gemini, and Pisces
Is where we hope to find a cloud of interstellar ices.
Orion, Piscis Austrinis, in Phoenix, and in Pegasus is
Where the clouds are dense enough and each then coalesces.

Toward Cassiopeia, toward the Giraffe, and also toward the Hunting Dogs
We find the stars and galaxies and scribble in our nightly logs.
Ursa Major/Minor, Vela, Virgo, Volans, and Vulpecula
Is where we plan to aim tonight with million-dollar specula.

To lofty mountains off we go in confident reliance,
In hot pursuit of photons there and world-beating science.

Endnotes

1. Allen, Richard Hinckley, *Star Names: Their Lore and Meaning* Mineola, New York: Dover, 1963 reprint of original 1899 edition.
2. Gurshtein, Alexander, "Did the Pre-Indo-Europeans Influence the Formation of the Western Zodiac?" *The Journal of Indo-European Studies*, 33, No. 1 & 2, Spring/Summer 2005, pp. 103–150.
3. http://www.ianridpath.com/startales/contents.htm (accessed May 11, 2013).
4. http://www.ianridpath.com/boundaries.htm (accessed May 11, 2013). Reproduced courtesy of Ian Ridpath.
5. http://www.southastrodel.com/Page207.htm (accessed May 11, 2013).
6. Tom Lehrer and Ronald Searle, *Too Many Songs by Tom Lehrer and Not Enough Drawings by Ronald Searle*, New York: Random House, 1981, pp. 63–65.

10

A Mathematical Sampler

This chapter is dedicated to the memory of Jaime Escalante (1930–2010). He was a high school math teacher who prepared his students for the Advanced Placement test in calculus and the protagonist of the film *Stand and Deliver* (1988) starring Edward James Olmos. Educational Testing Service of Princeton, New Jersey, did not believe that Escalante's students, who were poor Latinos from East Los Angeles, could pass this test without cheating. They were wrong.

Integration for People Who Don't Know Calculus

To become a scientist, one typically needs a mathematical grounding in algebra, geometry, trigonometry, graphing and analysis methods, differential and integral calculus, differential equations, and statistics. This is pretty intimidating for most people.

One standard purpose of integral calculus is to determine the area under a curve in a graph of some sort. This has countless applications in science and engineering. You might think you could never do integral calculus. But if you can use scissors, you can!

Without specifying the function in Figure 10.1, how would you determine the area under the curve (the thick black line at the top), let's say from $x = 1$ to $x = 5$? You could paste this graph onto a piece of thick cardboard, cut along the dashed lines, then along the thick black line at the top, then along the horizontal (x) axis. Weigh the cardboard very carefully (giving w_1). Then cut out one square unit and weigh that square (giving w_2). The area under the curve will be equal to the ratio w_1/w_2.

There is a second way to obtain the area under the curve using scissors. Cut it into a whole bunch of thin rectangular strips (of width, say, 0.1 of our x-axis units), as in Figure 10.2. Tape these strips together like one long fetuccini noodle of length L. Then the area will simply be

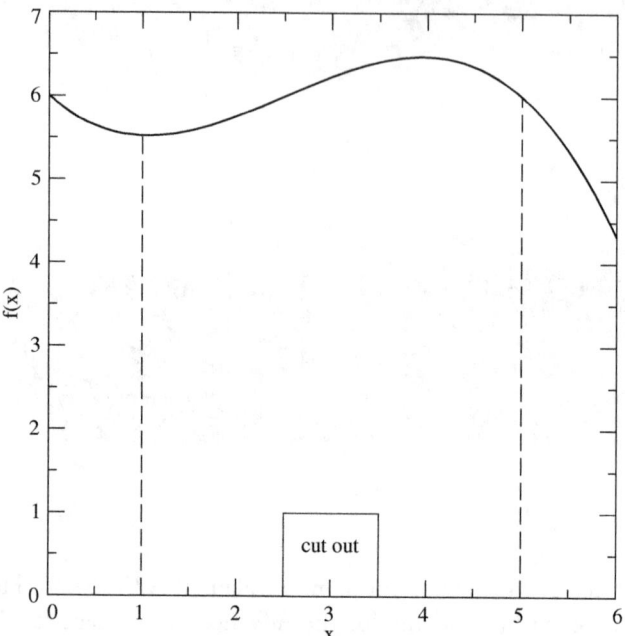

Figure 10.1 One way to find the area under the function f(x) is to paste this graph onto thick cardboard, cut out the desired area (e.g., between the dashed lines), weigh it, then weigh one square unit of the graph. The ratio will give the area under the curve in terms of those square units.

$L \times 0.1$. Why? Because the area of a rectangle is the length times the width, no matter how skinny the rectangle. This method has a name for mathematicians and computer programmers. It is called *numerical integration*.

Say we actually specify the function shown in Figures 10.1 and 10.2:

$$f(x) = 6 - x + 0.6x^2 - 0.08x^3. \tag{10.1}$$

Having this function allows us to calculate the y-axis value for any value of x. The area of each strip is the y-axis value times the width of the strip. Adding up all these individual areas gives us the total area. Using this method we obtain an estimate of the area under f(x) from $x = 1$ to $x = 5$ of 24.3432 square units.

Now, if you know integral calculus you can easily obtain the integral of Equation 10.1:

$$F(x) = 6x - 0.5x^2 + 0.2x^3 - 0.02x^4. \tag{10.2}$$

(Your calculus teacher would say at this point, "You forgot to add 'plus a constant'," but it drops out in a moment, so we will leave it out.) $F(x)$ evaluated at $x = 5$ is 30.0. $F(x)$ evaluated at $x = 1$ is 5.68. The area under the curve f(x) from $x = 1$ to $x = 5$ equals $30.0 - 5.68 = 24.32$,

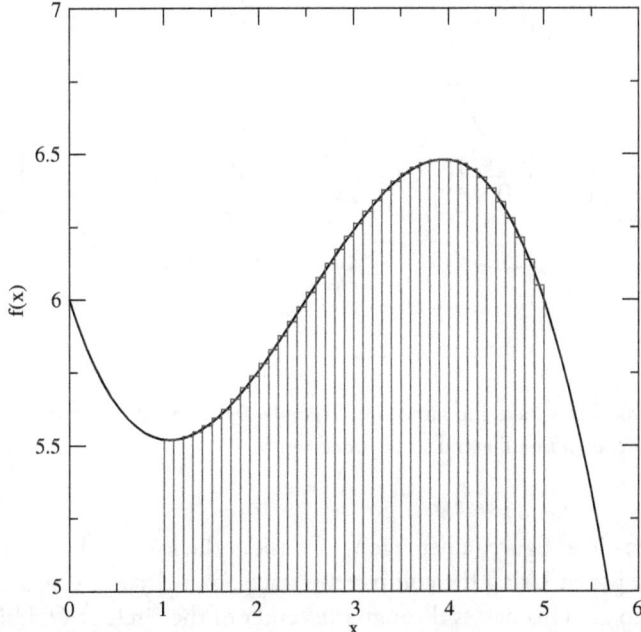

Figure 10.2 Imagine cutting the area shown in Figure 10.1 into a number of strips of width 0.1 units. By adding up the lengths of all the strips and multiplying by the width, we obtain the area under the curve from $x = 1$ to $x = 5$. We have zoomed in to show our approximation of the function that allows a simple numerical sum to give an accurate value of the area under the curve.

exactly. Thus, our result from numerical integration has a systematic error of 0.0232 square units, which amounts to 0.1 percent. We might get a more accurate answer from numerical integration by making the slices much thinner, but eventually we confront a problem known as *roundoff error*. There is an accuracy limit to our calculations using "single precision" numbers. We could convert our program to use "double precision" numbers, but the problem never completely goes away. Calculus often provides the exact answer. Yet there are some functions we *must* integrate numerically, if no explicit mathematical solution exists.

Drawing Figures with a Ruler and Compass

An equilateral polygon is a figure drawn on a flat piece of paper that has sides of equal length, and the interior angles are all the same as well. In my high school geometry class we were shown how to construct an equlilateral triangle, square, and hexagon using a compass and a ruler. More recently I found an example on the web showing how to construct an equilateral pentagon.[1]

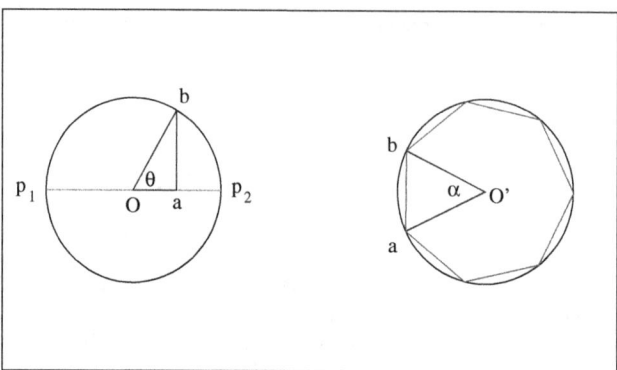

Figure 10.3 The two circles have the same size. Triangle Oab is a 30-60-90 degree right triangle. The chord ab allows us to create an equilateral seven-sided figure.

How about a seven-sided figure (a heptagon)? Consider the left-hand circle in Figure 10.3. We make a point on the paper. Using the compass we trace out a circle. Using a ruler we then draw a diameter from p_1 to p_2. This passes through the center of the circle at O. Using a compass and ruler we can bisect radius Op_2 at point a. Triangle Oab is a 30-60-90 degree right triangle. The length from a to b is equal to the radius of the circle times $\sqrt{3}/2$. The right-hand circle has the same radius as the left-hand circle and the chord ab has the same length. Using a compass we can then work our way around the right-hand circle and make an equilateral heptagon.

I discovered that this works out to within the width of a pencil line, but is it mathematically exact? If it is truly an equilateral heptagon, the chord ab in the right-hand circle will subtend an angle equal to $360/7 \approx 51.42857 \ldots$ degrees. It can be shown that angle α in the right-hand circle is equal to two times the angle whose trigonometric sine is equal to $\sqrt{3}/4 \approx 51.3178 \ldots$ degrees. This is about 0.1 degree less than the number we were hoping for. Thus, our method works pretty well, but is not exact.

Some Fun with Angles

In Figure 10.3 the trigonometric **S**ine of angle θ is equal to the length of the **O**pposite side (ab) divided by the **H**ypotenuse (r). The trigonometric **C**osine of angle θ is equal to the length of the **A**djacent side (Oa) divided by the **H**ypotenuse. The trigonometric **T**angent of angle θ is equal to the length of the **O**pposite side divided by the **A**djacent side. This gives us the mnemonic **SOH-CAH-TOA**, which can be memorized by most thirteen-year-olds.

On my calculator I have functions sine, cosine, and tangent, and also their inverse functions. For example, the angle whose sine is equal to $\sqrt{3}/2$ is sixty degrees. The notation would be $\sin^{-1}(\sqrt{3}/2) = 60°$. Here the exponent -1 does not mean a reciprocal, but the inverse function.

The algorithm shown in reference 1 to inscribe an equilateral pentagon in a circle allows us to prove two curious facts:

$$\sin^{-1}\left(\frac{\sqrt{5}-1}{4}\right) = 18 \text{ degrees, exactly.} \tag{10.3}$$

$$\sin^{-1}\left(\frac{\sqrt{5}+1}{4}\right) = 54 \text{ degrees, exactly.} \tag{10.4}$$

This got me to thinking, and I wrote a little program to look at fractions of the form $(\sqrt{I}+J)/K$, where I and K are positive integers and J is a positive integer, negative integer, or zero. Are any of these combinations equal to some angle other than multiples of thirty or forty-five degrees? Here is one:

$$\tan^{-1}(\sqrt{3}+2) = 75 \text{ degrees, exactly.} \tag{10.5}$$

This can be verified using the tangents of thirty and forty-five degrees and a trigonometric identity for the tangent of the sum of two angles.

But then I stumbled onto this:

$$\sin^{-1}\left(\frac{\sqrt{42}-4}{9}\right) = 16.00003 \text{ degrees.} \tag{10.6}$$

It is close, but not exact. This is undoubtedly known, as it is simple to derive. But it was found independently by me, and for that I feel some satisfaction. Nobody I have shown this to (which includes a number of Ph.D. astronomers and physicists) seems to have ever seen this approximation.

Three Mathematical Geniuses

Ramanujan

When I was growing up there was a set of four books on the shelf of my father's study with these intimidating symbols (Σ) on the spines. This was *The World of Mathematics*, edited by James R. Newman (1956). I just assumed that I wouldn't understand what was inside. But to my surprise, when I was twenty-one, a fellow graduate student recommended that I check them out. I found that these volumes were mostly sentences and paragraphs, not equations. Many of the articles were *about* mathematics and mathematicians.

It was in volume 1 of Newman's anthology that I first encountered Srinivasa Ramanujan, an Indian genius who lived from 1887 to 1920.[2,3,4] He was a self-taught mathematician who was so poor he could hardly afford paper. So he would work out a mathematical problem on a small chalkboard and copy the question and the answer into a notebook. But he would *not* copy over the derivation or proof. Mathematicians have spent the better part of a century trying to confirm Ramanujan's findings. His work was highly original. A lot is really intimidating.

But here's an example we can all understand. Once, when Ramanujan was in the hospital in England, he was visited by his friend and colleague, the mathematician G. H. Hardy. Hardy mentioned offhand that he had come in cab number 1729 and that he thought this was a rather dull number. Ramanujan said, "No, that is a very interesting number. It is the smallest number that is the sum of two cubes in two different ways." You should remember that the cube of a number is the number multipled by itself twice more. Thus, $9^3 = 9 \times 9 \times 9 = 729$. $10^3 = 1000$. $1^3 = 1$, and $12^3 = 1728$. What Ramanujan was saying was:

$$1729 = 9^3 + 10^3 = 1^3 + 12^3. \tag{10.7}$$

The number 1729 is referred to as the Hardy-Ramanujan number.

Hardy asked Ramanujan if he knew what was the smallest number that is the sum of two cubes in three different ways. Ramanujan didn't know. This was only discovered in 1957 by the British mathematician John Leech. It is:

$$87539319 = 167^3 + 436^3 = 228^3 + 423^3 = 255^3 + 414^3. \tag{10.8}$$

The numbers 1729 and 87539319 are known as "taxicab numbers." Larger ones have been found by computer.

In the April 1975 issue of *Scientific American*, Martin Gardner wrote about one of Ramanujan's independent discoveries. The special number e is the base of the natural logarithms, equal to 2.71828.... e^x is probably a function on your calculator. Ramanujan found that

$$e^{\pi\sqrt{163}} = 262,537,412,640,768,744 \text{ (exactly)}. \tag{10.9}$$

An eighteen-digit integer! How can this be? e and π ($= 3.14159265...$) are "transcendental numbers." (For the mathematically inclined, this means that they cannot be roots of algebraic equations with rational coefficients.) $\sqrt{163}$ is irrational. How could this function of nonrepeating decimal numbers be a whole number? This long number is known as Ramanujan's constant, and it turns out that it was actually first discovered by the French mathematician Charles Hermite in 1859. It also turns out that Martin Gardner was playing an April Fool's joke on us. The number is not an integer that ends with a 4. We must replace that final 4 with 3.999 999 999 999 25 ... So, Ramanujan's constant is *almost* an integer.

One of the sad things about Ramanujan was that he died at a young age, only thirty-three. Who knows what he would have accomplished had he lived to a ripe old age. But maybe it would not have been proportionately more original work. Most mathematicians do their best work by the time they are thirty-something.

> Note: The next two subsections may be intimidating if the reader has not studied advanced calculus. You may skip these and go make yourself a nice cup of tea.

Euler

One mathematician who was productive his whole life was Leonhard Euler (1707–1783). He was born in Switzerland but spent most of his adulthood in St. Petersburg (Russia) and Berlin. The most amazing thing about Euler is that from 1726 until 1800 his work comprised roughly one-third of all the mathematics papers published in the entire world! His collected works run to seventy-four volumes with seven volumes yet to be published.[5]

Euler proved directly that

$$e^x = \sum_{n=0}^{\infty} \frac{x^n}{n!} = \lim_{n \to \infty} \left(\frac{1}{0!} + \frac{x}{1!} + \frac{x^2}{2!} + \frac{x^3}{3!} + \ldots + \frac{x^n}{n!} \right). \tag{10.10}$$

Here 0! ("zero factorial") $= 1$, and $n! = n \times (n-1) \times (n-2) \ldots 1$. Previously, this had only been proven indirectly by Newton and Leibniz.

One of Euler's early significant results was the solution to the so-called Basel problem:

$$\sum_{n=1}^{\infty} \left(\frac{1}{n^2} \right) = \lim_{n \to \infty} \left(\frac{1}{1^2} + \frac{1}{2^2} + \frac{1}{3^2} + \ldots + \frac{1}{n^2} \right) = \frac{\pi^2}{6}. \tag{10.11}$$

The thing to appreciate here is that we have a specific answer related to the famous number π, not just an approximation.[6]

One way to confirm this is to use some mathematics from the nineteenth century called Fourier analysis.[7] We can represent the simple function $f(x) = x^2$ by an infinite sum of cosine terms of decreasing amplitude:

$$f(x) = x^2 = \frac{\pi^2}{3} + 4 \sum_{n=1}^{\infty} \frac{(-1)^n \cos(nx)}{n^2}. \tag{10.12}$$

We show this in Figure 10.4. Now let us set $x = \pi$ radians (equivalent to 180 degrees). Since $(-1)^n \cos(nx) = 1$ for any positive integer n, it follows that

$$\pi^2 = \frac{\pi^2}{3} + 4 \sum_{n=1}^{\infty} \frac{1}{n^2}. \tag{10.13}$$

This directly gives the result shown in Equation 10.11.

It is more challenging to prove that

$$\sum_{n=1}^{\infty} \left(\frac{1}{n^4} \right) = \frac{\pi^4}{90} \tag{10.14}$$

and

$$\sum_{n=1}^{\infty} \left(\frac{1}{n^6} \right) = \frac{\pi^6}{945}. \tag{10.15}$$

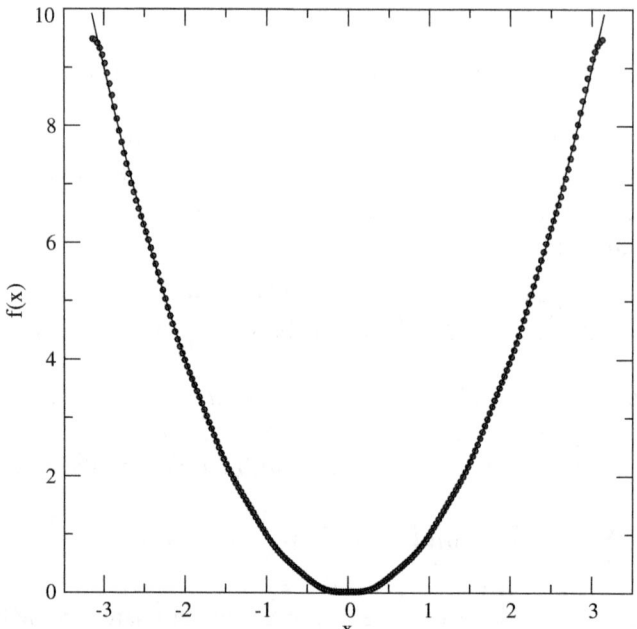

Figure 10.4 Here we have calculated the first ten cosine terms of Equation 10.12 and plotted the sums each 0.03 radians (the dots). The thin solid line is the actual function $f(x) = x^2$. Except near $x = \pm\pi$ the agreement is very good with only ten terms.

These can be worked out using something called Parseval's equation[8] and the Fourier series equivalents for $f(x) = x, x^2$, and x^3.

Mathematicians will recognize the results of Equations 10.11, 10.14, and 10.15 as values of the Riemann zeta function $\zeta(s)$ for $s = 2, 4$, and 6.[9]

Finally, the Riemann zeta function is related to the area under the curve of a function of the form:

$$f(x) = \frac{x^{s-1}}{e^x - 1}. \tag{10.16}$$

If $s = 4$, the integral of Equation 10.16 from $x = 0$ to $x = \infty$ is equal to 6 times the result from Equation 10.14, or $\pi^4/15$.[10,11]

What does this have to do with anything in the real world? Well, consider the similarity of Equation 10.16 (for $s = 4$) and the formula for the spectral radiance (B_ν) of a hot solid, liquid, or dense gas at temperature T:

$$B_\nu = \frac{2h}{c^2} \frac{\nu^3}{e^{\frac{h\nu}{kT}} - 1}, \tag{10.17}$$

where ν is the frequency of light, h is Planck's constant, c is the velocity of light, and k is Boltzmann's constant. What do we mean by "hot"? Any solid, liquid, or dense gas will give off light if it is hotter than absolute zero. So, it does not have to be really hot. The temperature of the Earth or a human is about three hundred deg K. Your body radiates light—essentially no optical light—but you do radiate infrared light, and the maximum intensity of that infrared light is at a wavelength of about ten microns, roughly twenty times the wavelength of light our eyes are most sensitive to. This is how infrared night goggles allow one soldier to see another soldier "in the dark."

Equation 10.17 allows us to calculate the total energy that the Sun (or any other star) emits over all frequencies (i.e., from low-frequency radio waves to high-frequency gamma rays). It also leads us to the result that the total energy given off by a star per square meter is proportional to the fourth power of the temperature. If the Sun's photosphere were to get 8 percent hotter, the Sun would give off $(1.08)^4 \approx 1.36$ times as much light. Imagine how that might affect life on Earth or at least the evolution of eyesight. Over the past 4.6 billion years the Sun *has* actually gotten about one-third brighter, but the temperature on the Earth has remained remarkably constant.[12]

Gauss

We would also like to shine the spotlight here on the mathematician and astronomer Carl Friedrich Gauss (1777–1855). His precociousness was noted at a young age. One day while in class his teacher told his students to add up one hundred numbers that made an arithmetic progression. A simple example would be to add up the first hundred whole numbers: $1 + 2 + 3 + \ldots 100$. In less than a minute Gauss had the answer. How did he do it? We learn in math class that the order of addition does not matter. In this case the sum is the same as $(1 + 100) + (2 + 99) + (3 + 98) \ldots$ We have fifty pairs of numbers, each pair of which adds up to 101. And $50 \times 101 = 5050$. Voila! This teacher of Gauss, rather than being irritated, realized he had a special pupil. He found him more advanced tutors. And Gauss went on to become one of the most significant mathematicians of the nineteenth century.[13]

Gauss is most famous for giving us the bell-shaped curve, which is applicable to so much of statistics (see Figure 10.5). Roughly 68.3 percent of the data that produces such a distribution falls within one standard deviation of the mean, 95.5 percent of the data falls within two standard deviations of the mean, and 99.7 percent of the data falls within three standard deviations of the mean. Often, data points that fall more than three standard deviations from the mean are termed *outliers* and are eliminated from the analysis.

Consider the function $f(x) = e^{-x^2}$. How would we find the area under the curve for the full range of x?

$$\int_{-\infty}^{\infty} e^{-x^2} dx = \text{some constant } k? \tag{10.18}$$

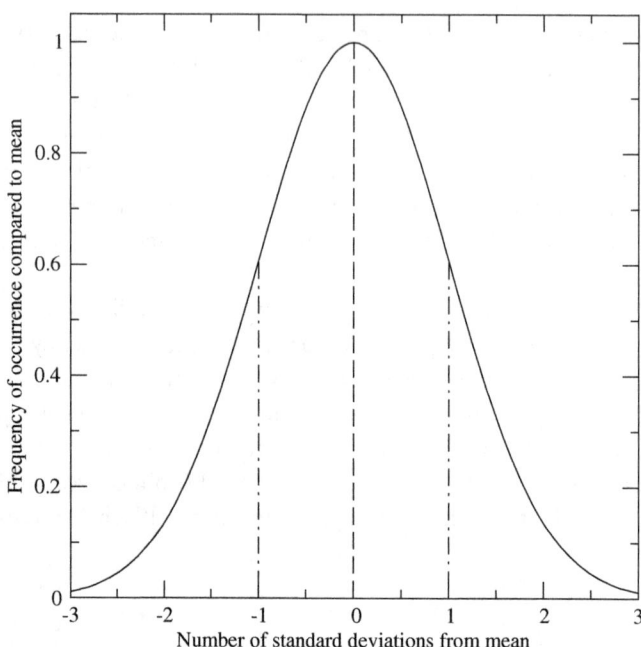

Figure 10.5 The normal curve of error. Roughly 68.3 percent of the data falls within one standard deviation of the mean (bounded by the two vertical dot-dashed lines). The area under the solid curve from $x = -\infty$ to 0, from 0 to $+\infty$ or from $x = -\infty$ to $x = +\infty$ can be calculated exactly, but otherwise the area must be determined via numerical integration.

No matter how we search for a substitution to make something easy to integrate, such as $e^u\, du$, we cannot. But we can hypothesize that the integral shown in Equation 10.18 is equal to some constant, say k. Then the following will also be true:

$$\int_{-\infty}^{\infty} e^{-y^2}\, dy = \text{the same constant } k. \tag{10.19}$$

We can then make a two-dimensional integral:

$$\int_{-\infty}^{\infty} \int_{-\infty}^{\infty} e^{-(x^2+y^2)}\, dx\, dy = k^2. \tag{10.20}$$

We can convert this to polar coordinates (r and θ) as follows:

$$\int_{0}^{\infty} \int_{0}^{2\pi} e^{-r^2} r\, dr\, d\theta = k^2. \tag{10.21}$$

This *can* be easily converted into something of the form $e^u\, du$, and we find that the constant $k = \sqrt{\pi}$.

The normal curve of error ("the Gaussian distribution") is usually given as follows:

$$f(x) = \frac{1}{\sigma\sqrt{2\pi}} e^{\frac{-(x-\mu)^2}{2\sigma^2}}. \tag{10.22}$$

Here μ is the mean value and σ is the standard deviation of the distribution. The scaling factor $1/(\sigma\sqrt{2\pi})$ makes the area under the curve equal to 1. (In Figure 10.5 the mean value is 0 and the standard deviation is 1.) So the probability of obtaining a result *from* some number of standard deviations with respect to the mean *to* some other number of standard deviations with respect to the mean is equal to the area under the curve using Equation 10.22. However, this result has to be obtained with numerical integration unless one is dealing with the left half, the right half, or all of the distribution shown in Figure 10.5.

Here is a tangible example. Say a golfer has an *average* score of 10.6 over par, and the standard deviation of his scores of ± 4.3 strokes. What is the probability that he would score between 0 and 7 over par (a round of 72 to 79)? It turns out that the probability is 0.2304. If this person played 506 rounds of golf over some number of years, Gaussian statistics would stipulate that he would have had 117 rounds in the 70s and 2 rounds of even par. And, sure enough, the golfer represented here actually had 112 rounds in the 70s and 2 rounds of even par.

If we find that some characteristic of a dataset is a function of some other characteristic, we can find a robust mathematical relation between the variables. The simplest relationship would be a linear one, like those shown in Figure 10.6. Gauss elaborated a method to determine the

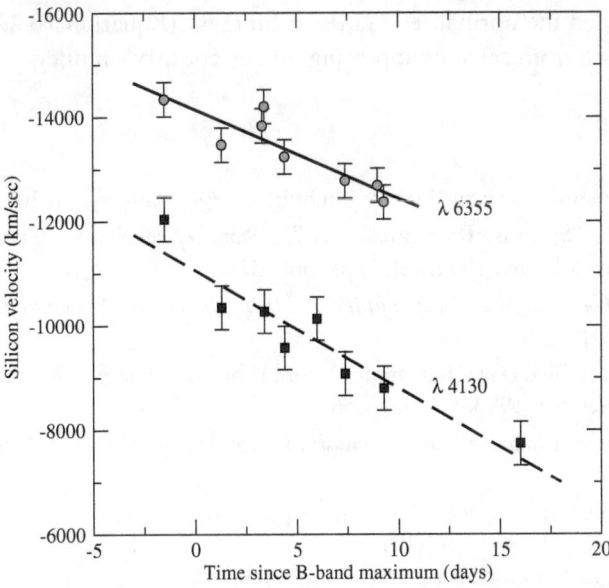

Figure 10.6 Example of two linear fits to data using Gauss's method of least-squares. From reference 14.

Figure 10.7 Gauss and the Gaussian distribution, as pictured on the German ten-Mark bill.

slope and intercept of the fit that minimizes the sum of squares of the deviations (i.e., differences above and below the line). These deviations will be distributed above and below the line according to a Gaussian distribution with some particular standard deviation.

Few countries have honored mathematicians on their paper currency. From 1978 to 1988 one-pound notes issued in Great Britain showed Isaac Newton. In Germany prior to the existence of the European Union and the euro, Gauss appeared on the ten-Mark note (see Figure 10.7). Not only that, they included the normalized Gaussian function (Equation 10.22), perhaps the only example of an equation from calculus appearing on any country's money.

Endnotes

1. http://www.mathopenref.com/constinpentagon.html (accessed January 24, 2013).
2. Newman, James R., "Srinivasa Ramanujan," in *The World of Mathematics*, James R. Newman, ed., New York: Simon & Schuster, 1956, vol. 1, pp. 368–376.
3. Kanigel, Robert, *The Man Who New Infinity: A Life of the Genius Ramanujan*, New York: Charles Scribner's Sons, 1991.
4. Berndt, Bruce C. & S. Bhargava, "Ramanujan—for lowbrows," *American Mathematical Monthly*, 100, No. 7, August–September, 1993, pp. 644–656.
5. Havil, Julian, *Gamma: Exploring Euler's Constant*, Princeton, New Jersey: Princeton University Press, 2003, p. xxii.
6. Alfred J. Lotka published a paper called, "The frequency distribution of scientific productivity," *J. of the Washington Acad. of Sciences*, 16 (12), pp. 317–323 (1926). In it he considered the number of chemists (N) listed in *Chemical Abstracts* who had published n = 1, 2, 3, … papers during the years 1907–16, finding a power law of the form $N = kn^{-\alpha}$, with $\alpha = 1.888$. For the lifetime publication totals of physicists for the entire range of history up to 1900, as listed in Auerbach's *Geschichtstafeln*

der Physik, he found a power law with index $\alpha = 2.021$. If we have a power law with index 2, then the fraction of authors who publish only one paper is the reciprocal of the result from Equation 10.11, or $6/\pi^2 \approx 0.608$. For astronomers the power law index is about 1.43. See Krisciunas, K., "Lotka's Law—Year by Year," *J. of the Amer. Soc. for Info. Science*, 28 (1), pp. 65–66 (1977).

7. Borowitz, Sidney, *Fundamentals of Quantum Mechanics*, New York and Amsterdam: W. A. Benjamin, Inc., 1967, p. 56, problem 3–10.
8. Kaplan, Wilfred, *Advanced Calculus*, Reading, Massachusetts: Addison-Wesley, 1952, pp. 412–413.
9. Derbyshire, John, *Prime Obsession: Bernhard Riemann and the Greatest Unsolved Problem in Mathematics*, Washington, DC: Joseph Henry Press, 2003.
10. Whittaker, E. T. & G. N. Watson, *A Course of Modern Analysis*, Cambridge: Cambridge University Press, 1935, pp. 243, 265–266.
11. Griffiths, David J., *Introduction to Quantum Mechanics*, Englewood Cliffs, New Jersey: Prentice Hall, 1995, pp. 216, 218.
12. Feulner, Georg, "The faint young Sun problem," *Reviews of Geophysics*, 50, RG2006 (also http://arxiv.org/abs/1204.4449).
13. Bell, Eric Temple, "The prince of mathematicians," in Newman, ref. 1, vol. 1, pp. 295–339.
14. Krisciunas, Kevin, et al.,"The most slowly declining Type Ia Supernova 2001ay," *Astronomical Journal*, 142, article 74 (September 2011).

11
Homework

On October 6, 1987, I spent half a day with Professor Subrahmanyan Chandrasekhar (1910–1995), the 1983 corecipient of the Nobel Prize in Physics. A transcript of my morning interview with him can be obtained from the American Institute of Physics.

Chandra is most famous for his derivation, at a young age, that the maximum mass of a white dwarf star is about 1.4 solar masses. This is called the "Chandrasekhar limit." The "Schönberg-Chandrasekhar limit" gives the maximum ratio of the mass of the core of a main sequence (hydrogen-burning) star (like the Sun) compared to the mass of the core plus envelope. This is between 0.10 and 0.15.[1]

Over lunch at the University of Chicago Faculty Club, I said to Chandra, "You must be pleased that there are a couple things named after you." He modestly replied, "That's fine, but fifty years down the road most science becomes anonymous. People will eventually just call it the white dwarf mass limit. Nobody calls it the Rutherford model of the atom anymore."

On April 7, 2010, I was at Cook's Branch Nature Conservancy, a private six-thousand-acre ranch near Montgomery, Texas, owned by the Mitchell family. There was a barbecue as part of a two-week workshop for string theorists, including Stephen Hawking, who has come for a number of years when his health has permitted. After dinner I was recalling this story about Chandra to some string theorists. We were within earshot of Hawking. Some minutes later his computer synthesizer spoke up, "The Dirac Equation will always be known as the Dirac Equation. I hope this will also be true with Hawking radiation."

It turns out, I think, that Chandra was wrong. There are many, many things that do not become anonymous within fifty years. In what follows, we do not attempt to provide an exhaustive list of constants, parameters, and such named after people. Instead, we aim to make a list of different things that are part of math, physics, and astronomy that have some name associated with them. Thus, for *limit* we only need mention Chandrasekhar, but I give two other examples as well. Your homework is to figure out what these things are. For some things more than one example is given.

- Boolean **algebra**.
- Cooley-Tukey **algorithm** for Fast Fourier Transforms. Also, Lomb-Scargle algorithm for finding frequencies in the power spectra of unequally spaced data.
- Littrow **angle**.
- Stirling's **approximation** to n factorial. Also, Sobolev approximation (radiative transfer).
- Kuiper **belt**, Van Allen belts, Gould belt.
- Malmquist **bias**.
- Higgs **boson**. Bosons are named after Bose. They obey Bose-Einstein statistics. Bosons have integral spin.
- Julian **calendar**, Gregorian calendar. These relate to why the day after October 4, 1582, was October 15.[2]
- Schroedinger's **cat**, which is a thought experiment about quantum mechanics.
- Messier **catalogue**.
- Markov **chain**.
- Oort **cloud**, Magellanic Clouds.
- Chebyshev **coefficients**.
- Tesla **coil**.
- Riemann **Conjecture** (famous).[3] Also, Merton-Trimble-Krisciunas Conjecture (obscure).[4]
- Planck's **constant**, Euler's constant, Boltzmann constant, Hubble constant. Also Hubble expansion, Hubble flow.
- Cartesian **coordinates**.
- Milankovitch **cycle**, Carnot cycle.
- Maxwell's **demon**.
- Hertzsprung-Russell **diagram**, Hubble diagram, Feynman diagram.
- Parenago's **discontinuity**, Mohorovičič discontinuity.
- Gaussian **distribution**, Poisson distribution.
- Mössbauer **effect**, Zeeman effect, Matthew Effect (which was coined by sociologist Robert Merton—it applies to scientists getting more credit than they may deserve).[5] Also, Lutz-Kelker effect, Sunyaev-Zel'dovich effect.
- Fermi **energy**, Fermi gas, Fermi temperature, Fermi-Dirac statistics. Fermions have half-integral spin.

- Carnot **engine**.
- Friedman **equation**, Dirac equation, Navier-Stokes equation, Maxwell's equations. There are many examples.
- Kellner **eyepiece** (also Erfle, Ramsden, Plössl, Nagler).
- Cassegrain **focus**, Newtonian focus, Nasmyth focus.
- Seward's **folly**. (OK, this had to do with US history.)
- Coriolis **force**.
- Riemann zeta **function**, Bessel function.
- Seyfert **galaxy**.
- Hertzsprung **gap**.
- Euclidean **geometry**.
- Bok **globule**.
- h-index, or Hirsch **index**.
- Michelson **interferometer**.
- **Islets** of Langerhans (not from geography!).
- Newton's **Law** of Gravity, Kepler's Laws of planetary motion, Kirchhoff's Laws for spectra, Wien's Law, Hubble's Law. . . .
- Planck **length**, Jeans length.
- Fabry **lens**.
- Chandrasekhar **limit**, Dawes limit, Eddington limit.
- Fraunhofer **lines**.
- Roche **lobe**.
- Jeans **mass**, Planck mass.
- Rosseland **mean opacity**.
- WKB **method**. The initials stand for Wentzel, Kramers, and Brillouin. Also, Newton-Raphson method.
- Einstein-de Sitter **model**.
- Brownian **motion**.
- Dobsonian **mounting**.
- Poisson **noise**.
- Avogadro's **number**.
- Schmidt-Cassegrain **optical system**. Also, Maksutov optical system.
- EPR **paradox**. EPR stands for Einstein-Podolsky-Rosen.

- Foucault **pendulum**.
- Inner Lagrangian **point**.
- Hermite **polynomials**.
- Pauli exclusion **principle**, Heisenberg Uncertainty Principle.
- Nobel **Prize**, which is why we remember Alfred Nobel, the discoverer of dynamite.
- Hawking **radiation**, Cerenkov radiation.
- Bohr **radius**, Schwarzschild radius.
- Strehl **ratio**.
- Occam's **razor**.
- Phillips **relation**, Faber-Jackson relation, Tully-Fisher relation (all astronomical).
- Ritchey-Chrétien **reflector**.
- Einstein **ring**, Newton's rings.
- l'Hôpital's **rule**.
- Richter **scale**, Beaufort scale.
- Mie **scattering**, Compton scattering, Rayleigh scattering.
- Taylor **series**, Fourier series, Balmer series of hydrogen lines.
- Bourne **shell**.
- Doppler **shift**, Lamb shift.
- Hilbert **space**, Banach space.
- Hubble **Space Telescope**.
- Fourier **spectrometer**.
- Dyson **sphere**.
- Barnard's **star**.
- Lyot **stop**.
- Pythagorean **theorem**, Fermat's Last Theorem.
- Planck **time**, Hubble time.
- Fourier **Transform**, Laplace Transform.
- Hubble **type**.
- Pascal's **wager**. (Philosophy and theology.)
- Alfvén **waves**.
- Baade's **window**.
- And last, but not least, the **witch** of Agnesi. Not a person. Has to do with mathematics.

Endnotes

1. Schönberg, M. & S. Chandrasekhar, 1942, "On the evolution of main-sequence stars," *Astrophysical Journal*, 96, pp. 161–172.
2. This has to do with the use of leap years to keep the calendar in sync with the vernal equinox. In 46 BC Julius Caesar decreed that every fourth year would be a leap year, which is almost but not quite right, as the year is 365.2422 days long. By the sixteenth century the calendar was ten days out of step with the sky. Pope Gregory XIII authorized the calendar reform named after him. In addition to the calendar adjustment, years divisible by four hundred (like 1600 and 2000) would be leap years, but years such as 1700, 1800, and 1900 would not. The British Empire converted to the Gregorian calendar in 1752, Russia in 1918, and Greece in 1923, by which point the adjustment was thirteen days. See: *Gregorian Reform of the Calendar: Proceedings of the Vatican Conference to Commemorate its 400th Anniversary, 1582–1982*, G. V. Coyne, M. A. Hoskin, and O. Pedersen, eds., Pontificia Academia Scientiarum, 1983.
3. Derbyshire, John, *Prime Obsession: Bernhard Riemann and the Greatest Unsolved Problem in Mathematics*, Washington, DC: Joseph Henry Press, 2003.
4. Trimble, Virginia & Markus J. Aschwanden, 2000, "Astrophysics in 1999," *Publications of the Astronomical Society of the Pacific*, 112, pp. 434–503, on p. 482. The idea of this conjecture is that something can be associated with three names but very rarely with four or more. The EPR paradox, for example, might be referred to differently if the key 1935 paper had had more than three authors. See also WKB method. Along these lines the Nobel Prize in Physics cannot be awarded to a group. It can be awarded to a maximum of three people in any given year, even though the results came about as a result of the work of one or more groups. In 2011, for example, Brian Schmidt, Adam Riess, and Saul Perlmutter received the Nobel Prize for work done by the High-Z Supernova Team and the Supernova Cosmology Project.
5. Gladwell, Malcolm, *Outliers: The Story of Success*, New York, Boston, London: Little, Brown and Co., 2008, p. 25.

12

A Breed Apart

One of the most curious volumes on my shelf is called *Drunken Goldfish and Other Irrelevant Scientific Research* by William Hartson.[1] The author's hobby is perusing scientific literature (mostly medical research) to discover some of the curious things that researchers have actually published. Early on he points us to pioneering research by Ryback,[2] in which goldfish were taught to turn left or right while swimming in a mildly alcoholic solution. Then some fish were placed in a high-alcohol solution; within an hour some looked sluggish, and some turned over on their sides. Other fish were spared the high-alcohol swim. Three days later all fish (now sober) were retested in the mildly alcoholic solution. "Those who had never tasted strong drink remembered their learning task excellently. Those who had blacked out forgot it, while those who had sampled the heavy liquor without collapsing also retained most of their learning. *So a mildly inebriated goldfish will remember what you teach it, unless it continues drinking until it is paralytic.*"

Even more curious is the result of three Harvard researchers on the spermicidal properties of different kinds of Coca-Cola.[3] They found that Diet Coke killed them dead, dead, dead. Hartson writes: "What the experimenters do not take into account, however, is whether sperm actually like Coke. Perhaps they drank a lot of the Diet Coke and slimmed themselves to death."[4]

However, Hong et al. (1987) dispute the findings of Umpierre et al. (1985):

> The inhibitory effect of Old Coke, caffeine-free New Coke, New Coke, Diet Coke and Pepsi-Cola on human sperm motility was studied with a trans-membrane migration method. None of them could decrease sperm motility to less than 70 percent of control within one hour. A previous study which claimed a marked variation of spermicidal potencies among different formulations of Coca-Cola could not be confirmed. Even if cola has a spermicidal effect, its potency is relatively weak as compared with other well-known spermicidal agents.[5]

• • • • • • • •

Speaking of sperm cells, there is a species of fruit flies (*Drosophila bifurca*) that produces sperm cells that are twenty times longer than the adult fruit flies.[6] Imagine if this were true for humans. The sperm would be 115 feet long.

• • • • • • • •

According to three researchers at the University of Padua, "When it comes to mating, male guppies treasure their ugly friends, which make them look good by comparison." Also, the uglier the male guppy, the less likely it was that he would hang around brightly colored male fish, which are considered more handsome by female guppies.[7] From the original paper we read:

> Recent theory predicts that in species where females tend to mate with the relatively most ornamented males, males may increase their attractiveness to females, and hence mating success, by preferentially associating with females that are surrounded by less ornamented competitors. Despite this prediction, we still lack explicit experimental evidence that males strategically prefer females surrounded by less attractive competitors to maximize their relative attractiveness. In this paper, we provide a comprehensive test of this hypothesis. . . .

• • • • • • • •

In the debate between evolution and intelligent design, consider some really bad design examples.[8] Why does a human have an appendix? You can live without it, and having one, if it bursts, can kill you. Why do men have nipples? They serve no purpose. How about the laryngeal nerve, which connects the brain to the larynx in mammals. Does it go there directly? No. It extends to the chest, loops around a lung ligament, and runs back up the neck to the larynx. In the giraffe that means the nerve must run twenty feet *extra* to get from the brain to the larynx. Not very intelligent design!

• • • • • • • •

One of the most brutal conquerors of human history was Genghis Khan (1162–1227). He and his grandsons (mostly) created the largest empire the world has ever seen.[9] From an analysis of the Y chromosomes of men over a large swath of Asia, it has been shown that 8 percent can be linked back to Genghis Khan. This amounts to 0.5 percent of the world's male population. There are sixteen million male descendants of Genghis Khan.[10]

In the Bible we learn in I Kings, chapter 11, verse 3, that King Solomon had some seven hundred wives and three hundred concubines. How many children he fathered is not known. *Based on actual evidence*, it appears that the most prolific parent in world history was Moulay Ismail the Bloodthirsty, aka Ismail ibn Sharif, who ruled Morocco from 1672 to 1727. He is alleged to have fathered 888 children.[11] We do not consider prolific contributors to modern sperm banks.

Now isn't this curious? The meek are supposed to inherit the Earth, but there are sixteen million male descendants of one of the most brutal conquerors in world history, and Moulay Ismail the Bloodthirsty had nearly nine hundred children.

• • • • • • • •

Are cats better than dogs? Well, one thing cats have going for them is that if a cat falls out of a building, it has a high likelihood of survival. When I first heard of this it sounded like there was an evil band of researchers dropping pets out of windows at different heights to gauge the injuries. But no. It was just that there were many cases of pets that fell and survived.[12] There is even a name for this: *high-rise syndrome*. Whitney and Mehlhaff (1987) report:

> High-rise syndrome was diagnosed in 132 cats over a 5-month period. The mean age of the cats was 2.7 years. Ninety percent of the cats had some form of thoracic trauma. Of these, 68% had pulmonary contusions and 63% had pneumothorax. Abnormal respiratory patterns were evident clinically in 55%. Other common clinical findings included facial trauma (57%), limb fractures (39%), shock (24%), traumatic luxations (18%), hard palate fractures (17%), hypothermia (17%), and dental fractures (17%). Emergency (life-sustaining) treatment, primarily because of thoracic trauma and shock, was required in 37% of the cats. Nonemergency treatment was required in an additional 30%. The remaining 30% were observed, but did not require treatment. Ninety percent of the treated cats survived.

Surprisingly, these researchers found that cats that fall from six stories and higher (that survive) have less severe injuries than cats that fall from two to five stories. It is hypothesized that cats have the ability to relax on the way down. What is the consequence? They manage to flatten themselves out and maximize their cross-sectional area, thus lowering their terminal velocity. This in turn lowers the kinetic energy at impact, leading to less severe injuries. And landing with maximum area minimizes the pressure on each square inch of the cat's body.

However, Vnuk et al. (2004) found that, "Falls [of cats] from the seventh or higher stories, are associated with *more* severe injuries [my emphasis] and with a higher incidence of thoracic trauma."[13]

Not to be outdone, Gordon et al. (1993) investigated dogs falling one to six stories. In eighty out of eighty-one cases the dog survived.[14] Apparently, in falls up to six stories cats and dogs survive with comparable percentages, but cats have the advantage for higher falls.

• • • • • • • •

Konrad Lorenz (1903–1989) was an Austrian zoologist, ethologist, and ornithologist. In 1966 he wrote *On Aggression*, which deals primarily with the aggressive behavior of geese and tropical fish. In chapter 10, Lorenz turns to rat behavior. Rats have a very strong sense of community association. Consider this:

> What rats do when a member of a strange rat clan enters their territory or is put in there by a human experimenter is one of the most horrible and repulsive things which can be observed in animals. The strange rat may run around for minutes on end without having any idea of the terrible fate awaiting it; and the resident rats may continue for an equally long time with their ordinary affairs till finally the stranger comes close enough to one of them for it to get wind of the intruder. This information is transmitted like an electric shock through the resident rat, and at once the whole colony is alarmed by a process of mood transmission which is communicated in the Brown Rat by expression movements but in the House Rat by a sharp, shrill, satanic cry which is taken up by all members of the tribe within earshot....
>
> Only rarely does one see an animal in such desperation and panic, so conscious of the inevitability of a terrible death, as a rat which is about to be slain by rats. It ceases to defend itself.[15]

In the final chapter of his book Lorenz tries to sound an optimistic note about diminished aggression in human society. But I got the impression that too often the behavior of humans is like that of rats. How long have they been fighting in the Middle East? On the other hand, Northern Ireland is in a much better state than it was twenty-five years ago.

• • • • • • • • •

In Robert Wright's book *The Moral Animal* (which is discussed further in chapter 18) we are reminded that in many species females are coy and males are not. In fact, many males are so singularly focused on mating that they will pursue things other than females of their species. Some kinds of male frogs make a specific "release call" if another male is making a homosexual advance. Male snakes might waste considerable time with a dead female. Male turkeys win first prize for conjugal confusion. They might court a stuffed replica of a female. OK, fine. But others become aroused by a female turkey's head suspended fifteen inches above the ground. Imagine his confusion when such a turkey goes to mount the female and there is no body. Male turkeys with even more imagination will find a wooden head sufficient, with or without eyes or a beak.[16]

• • • • • • • • •

In *The Tears of the Cheetah* geneticist Stephen O'Brien shows just how endangered many species are. One of the contributing factors is called a *bottleneck event*. If the population of a species (such as cheetahs) reaches a low enough level, the inbreeding and consequent passing on of mutations severely threatens the species.

To me the most interesting part of this book had to do with one consequence of the Black Death of the mid-fourteenth century.[17] At least 30 percent of Europe's population died from the plague. This amounted to many tens of millions of people. However, some descendants of the survivors have a distinct advantage when it comes to the human immunodeficiency virus (HIV), the virus that causes AIDS. It all has to do with a gene variant known as CCR5-Δ32. People who inherit *two* copies of this, one from each parent, are completely resistant to HIV infection, even if exposed to it again and again. The highest CCR5-Δ32 frequency is found in Scandinavia, Finland, and northern Russia (16 percent). In France, England, and Germany the rate is 10 percent. In Italy, Turkey, and Bulgaria it is 5 percent. In Saudi Arabia and sub-Saharan Africa it is 0 percent.[18]

• • • • • • • •

I enjoyed reading a book by Ed Regis called *Who Got Einstein's Office: Eccentricity and Genius at the Institute for Advanced Study* (1988). Related to that is this question: after Einstein died, what happened to his brain? It now resides in the National Institute of Health and Medicine, but it was in the possession of one Dr. Thomas Stoltz Harvey (and his heirs) from the time of Einstein's death in 1955 until 2010.[19]

Endnotes

1. Hartson, William, *Drunken Goldfish and Other Irrelevant Scientific Research*, New York: Sterling Publishing Co.(1988), pp. 7–9.
2. Ryback, R. S., 1969, "The use of goldfish as a model for alcohol amnesia in man," *Quarterly J. Studies on Alcohol*, 30, pp. 877–882.
3. Umpierre, S. A., J. A. Hill & D. J. Anderson, 1985, "Effect of Coke on sperm motility," *New England Journal of Medicine*, 313, p. 1351.
4. Hartson, ref. 1, pp. 61–63.
5. Hong, C. Y., C. C. Shieh, P. Wu & P. N. Chiang, 1987, "The Spermicidal Potency of Coca-Cola and Pepsi-Cola," *Human and Experimental Toxicology*, 6, pp. 395–396.
6. Pitnick, S., G. S. Spicer & T. A. Markow, 1995, "How long is a giant sperm?" *Nature*, 375, p. 109.
7. "Study shows ugly truth of guppy love," *Houston Chronicle*, February 17, 2013, p. A28. The original research was published in: Gasparini, Clelia, Giovanna Serena & Andrea Pilastro, "Do unattractive friends make you look better? Context-dependent male mating preferences in the guppy," *Proceedings of the Royal Society, B*, 280, No. 1756, 7 April 2013. http://rspb.royalsocietypublishing.org/content/280/1756/20123072.abstract
8. Holt, Jim, "Unintelligent design," *New York Times* magazine, February 20, 2005, p. 15.
9. Spuler, Bertold, *History of the Mongols*, New York: Dorset Press, 1988.
10. Zerjal, Tatiana, et al., 2003, "The genetic legacy of the Mongols," *Amer. J. Human Genetics*, 72, pp. 717–721.
11. Wright, Robert, *The Moral Animal: Evolutionary Psychology and Everyday Life*, New York: Vintage, 1994, p. 247.

12. Whitney, W. O. & C. J. Mehlhaff, 1987, "High-rise syndrome in cats," *Journal of the American Veterinary Medical Association*, 191 (11), pp. 1399–1403. Erratum: *Journal Amer. Vet. Med. Assoc.*, 1988 Feb 15, 192 (4), p. 542.
13. Vnuk, D., B. Pirkić, D. Maticić, B. Radisić, M. Stejskal, T. Babić, M. Kreszinger & N. Lemo, 2004, "Feline high-rise syndrome: 119 cases (1998–2001)," *J. Feline Med. Surg.* 6(5), pp. 305–312.
14. Gordon, L. E., C. Thacher & A. Kapatkin, 1993, "High-rise syndrome in dogs: 81 cases (1985–1991)," *J. Amer. Vet. Med. Assoc.*, 202(1), pp. 118–22.
15. Lorenz, Konrad, *On Aggression*, translated by Marjorie Kerr Wilson, San Diego, New York, London: Harcourt, Brace & Co., 1966, pp. 161–163.
16. Wright, ref. 11, pp. 46–47.
17. Cantor, Norman F., *In the Wake of the Plague: The Black Death and the World It Made*, New York: Free Press, 2001.
18. O'Brien, Stephen J., *Tears of the Cheetah and Other Tales from the Genetic Frontier*, New York: Dunne Books/St. Martin's Press, 2003, pp. 199, 211, 235–236.
19. http://en.wikipedia.org/wiki/Albert_Einstein's_brain (accessed March 14, 2013).

13

How to Get People to Tell the Truth

How do you get people to tell the truth when they don't want to? Say you were the minister of health of France, Great Britain, or any other country. You want to slow the spread of HIV (human immunodeficiency virus) in the population. In order to come up with the best plan, you have to know how people actually behave in your country. Never mind how you *think* they behave. You must devise a set of questions as neutral as possible and get a large number of people to answer these questions.

Typical public opinion polls involve asking questions of roughly one thousand people. This can be done by a polling company over the course of a weekend. On TV they often give such results and have a disclaimer that the results are accurate to ±3 percent. Where does that uncertainty come from? Statisticians tell me that it is somewhat complex, but from the astronomer's point of view it sounds just like photons from some distant quasar coming into your telescope. If you want to measure the brightness to ±3 percent, you have to detect one thousand photons above the brightness of the background sky. If N_{net} is the net number of photons, then the relative accuracy of your measurement is just $1/\sqrt{N_{net}}$. This is called *Poisson statistics*, but we are not going to get into that here.

There was an extensive survey of the sexual habits of 20,055 people in France and 18,876 people in Great Britain published in *Nature* in 1992, when HIV (the virus that causes AIDS) was effectively a death sentence.[1,2] These are huge numbers for such surveys. But one question you might ask yourself is this: How do we know that the people were responding honestly? If a man were ostensibly heterosexual, but occasionally had sex with another man, he might not want to admit it.

So let us have a short aside here and discuss a result described by Steven Levitt and Stephen Dubner in their popular book *Freakonomics*.[3] They wondered if standardized test scores in the city of Chicago were honest. What they found was that some of the teachers were cheating on behalf of the students. What was the motivation? If your students do well on the tests, that helps them get into college. And you may be rewarded for your students' good scores with a bonus or promotion. Uh, oh. This is a temptation, isn't it? What Levitt and Dubner found was that some teachers were taking the machine-gradable answer sheets, erasing *all* the answers for some fraction of the test, and replacing them with the correct answers. This is guaranteed to raise the average grade of the class.

The database was huge: seven hundred thousand sets of test answers. If the test is of an appropriate level of difficulty, some students will get a high percentage of right answers throughout the test, while others will get a much lower percentage of right answers. Having taught college full time for seven years, I see that there are some questions that are answered correctly by 90 percent of the students, while there are other questions that are answered correctly by only 40 percent. So, what is the probability that all of my students will get all of the questions right on the last quarter of the test? As high as the probability that monkeys writing at typewriters will never make so much as a typo error while reproducing all sorts of classic novels.[4] The net result was that teachers who were cheating for their students were discovered and fired.

In the fall of 2009 my colleagues and I in the Department of Physics and Astronomy at Texas A&M University were introducing "clickers" in our classrooms. The teacher can display a question via PowerPoint and the students can choose an answer (A through E) with a device that looks like a simple TV remote. You have instant feedback that avoids the embarrassment of calling on any individual student the old-fashioned way. Of course, the teacher's computer records the serial numbers of the clickers that were used for voting. This is done via a device from the clicker company that connects to the teacher's laptop. Using this system you can have many low-stress quiz questions. The students are usually allowed to talk amongst themselves before the voting times out. And there is a record of who was in class on a given day.

Here was a situation I found myself in, the day before Thanksgiving of 2009. The new university president had not cancelled classes as expected. I was lecturing twenty students of the seventy-five or so who were registered for my morning class. I displayed a clicker question and there were twenty-two answers. So I said, "OK. It seems that there are two more clickers than people in the class. Do not use your friend's clicker to garner him or her extra points. That's cheating." There were twenty answers on the next clicker question. It was easy to figure out who the phantom clickers belonged to.

You may be thinking that this is not a big deal, right? But not long before, twenty-four students in the Mays Business School at Texas A&M were caught cheating on a quiz.[5] This made national TV news. It was a class in business ethics!

Over the Thanksgiving weekend I read through Levitt and Dubner's second book, *Super Freakonomics*,[6] and I got inspired to do an experiment of my own the following Monday. That day, instead of twenty students present for my morning class, I had sixty-two. I asked them to get out their clickers and a coin. Sixty-one flipped a coin. Twenty-nine got heads and thirty-two got tails. If they all had "fair" coins, statistical theory stipulates that 68 percent of the time we should have gotten somewhere between 26.6 and 34.4 heads (call it twenty-seven to thirty-four) out of sixty-one. So far, so good.

Now the fun begins. I put on a sleeping mask, cupped my hands at the left and right edges, and faced away from the class. I instructed them to either keep their clicker or trade it with a friend. Thus, while a particular clicker might register a particular vote, I would not know if the owner of the clicker made that vote or another person did. Thus, we could assure that any given vote could not be traced back to a particular person with certainty.

"Now flip your coin again," I said. And I gave the following instructions:

- If you have ever loaned your clicker to another person to garner extra clicker points, and you flipped heads, click on A.
- If you have never loaned your clicker to another person to garner extra clicker points, click on B.
- If you *have* loaned your clicker to another person, but you flipped tails, I want you to lie and claim you never loaned your clicker to another person by answering B.

With this algorithm, not only would I not know who had whose clicker, but I would not know amongst the actual cheaters who was telling the truth and who was lying by asserting the opposite. If everyone abides by these rules that guarantee anonymity, then *twice* the percentage of people who clicked A is the best estimate of the fraction of people who have loaned a clicker to garner extra clicker points.

Here were the results. Out of a total of sixty-two students, eleven answered A, and fifty-one answered B. Thus, the best estimate of the fraction of cheaters was $(2 \times 11)/62 = 35$ percent. I have told my students since then about this experiment. And with a combination of clicker questions and a written quiz it is also possible to root out cheaters. So I have had minimal evidence of clicker cheaters since then.

Ethics in the classroom and in life in general is a good thing. That is why many universities, including ours, have an honor code. But it is not a life-and-death matter. The spread of HIV, on the other hand, *is*. Nowadays, if you live in a rich country, you may have access to an array of drugs that can keep you alive for many years, even though you are HIV positive. But in poor countries, that is not necessarily the case. The ministers of health of these countries need to know how people behave sexually to come up with the best plans to slow the spread of HIV. For that reason we need to employ testing algorithms like my clicker experiment in order to find out the truth from people. The respondents will want to feel very secure that an answer will be kept anonymous, and they may even be allowed to lie about what they did or did not do. If everyone being questioned abides by such a testing algorithm, the results will be honest and robust.

Endnotes

1. ACSF investigators, "AIDS and sexual behaviour in France," *Nature*, 360, pp. 407–409 (1992).
2. Johnson, M. Anne, Jane Wadsworth, Kaye Wellings, Sally Bradshaw & Julia Field, "Sexual lifestyles and HIV risk," *Nature*, 360, pp. 410–412 (1992).

3. Levitt, D. Steven & Stephen J. Dubner, *Freakonomics: A Rogue Economist Explores the Hidden Side of Everything*, New York: Harper Collins, 2005, pp. 25–37.
4. Maloney, Russell, "Inflexible Logic," in *The World of Mathematics*, James R. Newman, ed., New York: Simon & Schuster, 1956, vol. 4, pp. 2262–2267.
5. http://tamunews.tamu.edu/mays-takes-the-aggie-honor-code-seriously/ (accessed September 9, 2013).
6. Levitt, D. Steven & Stephen J. Dubner, *Super Freakonomics: Global Cooling, Patriotic Prostitutes and Why Suicide Bombers Should Buy Life Insurance*, New York: Harper Collins, 2009.

14

Famous First Words

One day in 2004, as I recall, I was listening to an interview on National Public Radio with a librarian from Seattle named Nancy Pearl. She gave ten or so examples of great openings of novels—something to grab your attention and make you want to read the whole book. For example, *The Towers of Trebizond* (1956) by Rose Macaulay begins, "'Take my camel, dear,' said my Aunt Dot, as she climbed down from this animal on her return from High Mass."

If you are a reader, or maybe even if you're not a reader, and you are looking for a good book to read, the very best places to look are Nancy Pearl's volume *Book Lust* and the sequel, *More Book Lust*.[1,2] There are 173 categories in the first book and 123 more in the second. Serious books, absurd books, fiction, nonfiction. You name it. I swear, this woman must have read more books than anyone since Theodore Roosevelt. Who knows? Maybe five thousand. I myself have probably not read one thousand yet. One short chapter in *Book Lust* is called "First lines to remember." And from that I am inspired to present a short list of my own here.[3]

Probably the most well-known opening lines are from *A Tale of Two Cities* by Charles Dickens (1859): "It was the best of times. It was the worst of times." Almost as well-known is the opening of Leo Tolstoy's *Anna Karenina* (1873 to 1877): "All happy families all alike. Every unhappy family is unhappy in its own way."

Perhaps in third place is the opening of *The Metamorphosis* by Franz Kafka (1915): *Als Gregor Samsa eines Morgens aus unruhigen Träumen erwachte, fand er sich in seinem Bett zu einem ungeheuren Ungeziefer verwandelt.* Translation: "One morning, as Gregor Samsa was waking up from anxious dreams, he discovered that in bed he had been changed into a monstrous verminous bug." The word Kafka used at the end of the opening sentence (*Ungeziefer*) means "vermin" but from context we can tell that poor Gregor had become an insect, not a rat.

In her radio interview, Nancy Pearl quoted the opening of *One Hundred Years of Solitude* (1967), the most famous book by Gabriel Garcia Marquez: "Many years later, as he was being led

Figure 14.1 The great Russian author Leo Tolstoy (1828–1910), author of *War and Peace* and *Anna Karenina*. Engraved by T[homas] Johnson (1887); photogravure by Sherer, Nabgoltz, and Company, Moscow. Private collection of Kevin Krisciunas.

to the firing squad, Colonel Aureliano Buendía was to remember that distant afternoon when his father took him to discover ice." Prior to reading that I read *Love in the Time of Cholera* (1985) by the same author and was tremendously impressed with his writing skill. It begins:

> It was inevitable: the scent of bitter almonds always reminded him of the fate of unrequited love. Dr. Juvenal Urbino noticed it as soon as he entered the still darkened house where he had hurried on an urgent call to attend a case that for him had lost all urgency many years before. The Antillean refugee Jeremiah de Saint-Amour, disabled war veteran, photographer of children, and his most sympathetic opponent in chess, had escaped the torments of memory with the aromatic fumes of gold cyanide.

Two of the main characters are young and in love, but the young woman marries a doctor and lives a comfortable life, while the young man goes on to a life akin to that of *Don Giovanni* from Mozart's opera about a libertine with a very long list of conquests. At the end of Garcia Marquez's novel, however, our two once-young lovers, now old and wrinkled, are together at last.

Moby Dick, by Herman Melville (1851), begins, "Call me Ishmael." As an astronomer who has worked on astronomical instrumentation, I like the quote from p. 544, where Captain Ahab dashes the astronomical quadrant to the deck, breaking it into pieces. Seething with contempt, he says, "Science! Curse thee, thou vain toy; and cursed be all things that cast man's eyes aloft to that heaven, whose live vividness but scorches him, as these old eyes are even now scorched with thy light, O Sun!"[4]

In the Soviet Union, officially, there was no God. This also meant that there was no devil. This makes the devil irritated, so he decides to come to Moscow along with a retinue that includes a naked girl vampire with red hair and phosphorescent eyes and a black cat that smokes cigars and is a dead shot with a revolver. On the second page of *The Master and Margarita* by Mikhail Bulgakov (1967) we read:

> Just then the sultry air coagulated and wove itself into the shape of a man—a transparent man of the strangest appearance. On his small head was a jockey cap, and he wore a short check jacket fabricated of air. The man was seven feet tall but narrow in the shoulders, incredibly thin and with a face made for derision ... the tall, transparent gentleman was swaying from left to right in front of [Berlioz] without touching the ground.

A parallel story relates to The Master, a Russian novelist of genius who is imprisoned for retelling the story of the Passion of the Christ in a way that is unpleasing to the Soviet authorities. He is a character reminiscent of Johann Faust, the conjurer of the devil in the German legend made famous by Christopher Marlowe (1604) and Johann Wolfgang von Goethe (1808).

Vanity Fair by William Makepeace Thackeray (1847–48) begins:

> While the present century was in its teens, and on one sunshiny morning in June, there drove up to the great iron gate of Miss Pinkerton's academy for young ladies, on Chiswick Mall, a large family coach, with two fat horses in blazing harness, driven by a fat coachman in a three-cornered hat and wig, at the rate of four miles an hour.

Robert Louis Stevenson's *Treasure Island* (1883) begins:

> Squire Trelawney, Dr. Livesey, and the rest of these gentlemen having asked me to write down the whole particulars about Treasure Island, from the beginning to the end, keeping nothing back but the bearings of the island, and that only because there is still treasure not yet lifted, I take up my pen in the year of grace 17__ and go back to the time when my father kept the Admiral Benbow inn and the brown old seaman with the sabre cut first took up his lodging under our roof.

From *The Sign of Four* by Sir Arthur Conan Doyle (1890):

> Sherlock Holmes took his bottle from the corner of the mantel-piece and his hypodermic syringe from its neat morocco case. With his long, white, nervous fingers he adjusted the delicate needle, and rolled back his left shirt-cuff. For some little time his eyes rested thoughtfully upon the sinewy forearm and wrist all dotted and scarred with innumerable puncture-marks. Finally he thrust the sharp point home, pressed down the tiny piston, and sank back into the velvet-lined arm-chair with a long sigh of satisfaction.

The most famous quote from this book is this: "Once you have eliminated the impossible, whatever remains, however improbable, must be the truth."

The Call of the Wild by Jack London (1903) begins:

> Buck did not read the newspapers, or he would have known that trouble was brewing, not alone for himself, but for every tidewater dog, strong of muscle and with warm, long hair, from Puget Sound to San Diego. Because men, groping in the Arctic darkness, had found a yellow metal, and because steamship and transportation companies were booming the find, thousands of men were rushing into the Northland. These men wanted dogs, and the dogs they wanted were heavy dogs, with strong muscles by which to toil, and furry coats to protect them from the frost.

The Offshore Pirate is a short story written by F. Scott Fitzgerald in 1920 that starts:

> This unlikely story begins on a sea that was a blue dream, as colorful as blue-silk stockings, and beneath a sky as blue as the irises of children's eyes. From the western half of the sky the Sun was shying little golden disks at the sea—if you gazed intently enough you could see them skip from wave tip to wave tip until they joined a broad collar of golden coin that was collecting half a mile out and would eventually be a dazzling sunset.

The dystopian novel *1984* by George Orwell (published in 1949) begins: "It was a bright cold day in April, and the clocks were striking thirteen."

The Magus by John Fowles (1966) starts out: "I was born in 1927, the only child of middle-class parents, both English, and themselves born in the grotesquely elongated shadow, which they never rose sufficiently above history to leave, of that monstrous dwarf Queen Victoria."

The opening of *Fear of Flying* by Erica Jong (1973): "There were 117 psychoanalysts on the Pan Am flight to Vienna and I'd been treated by at least six of them."

Speaker for the Dead by Orson Scott Card (1986) is a science-fiction novel that takes place about the year 5270. It begins:

> Since we are not yet fully comfortable with the idea that people from the next village are as human as ourselves, it is presumptuous in the extreme to suppose we could ever look at sociable, tool-making creatures who arose from other evolutionary paths and see not beasts but brothers, not rivals by fellow pilgrims journeying to the shrine of intelligence. Yet that is what I see, or yearn to see.

The Nutmeg of Consolation by Patrick O'Brian (1991) begins:

> A hundred and fifty-seven castaways on a desert island in the South China Sea, the survivors of the wreck of HMS Diane, which had struck upon an uncharted rock and had there been shattered by a great typhoon some days later: a hundred and fifty-seven men, but as they sat there round the edge of a flat bare piece of ground between high-water

mark and the beginning of the forest they sounded like the full complement of a ship of the line, for this was Sunday afternoon, and the starboard watch, headed by Captain Aubrey, was engaged in a cricket-match against the Marines, under their commanding officer, Mr. Welby.

The God of Animals (2007) by Aryn Kyle begins:

Six months before Polly Cain drowned in the canal, my sister, Nona, ran off and married a cowboy. My father said there was a time when he would have been able to stop her, and I wasn't sure if he meant a time in our lives when she would have listened to him, or a time in history when the Desert Valley Sheriff's Posse would have been allowed to chase after her with torches and drag her back to our house by her yellow hair.

"One day you know more dead people than live ones." From *Citizen Vince* by Jess Walter (2008). This novel is about a low-level mob soldier who is in the witness protection program and eventually comes out of hiding.

Biggest Elvis by P. F. Kluge (1997) is a laugh-out-loud novel about two brothers and a third-rate college professor from Guam who put together a nightclub act. The three represent the three stages of the career of Elvis Presley. There is Baby Elvis, who made the big splash, Dude Elvis, who made all those forgettable movies, and Biggest Elvis, the bloated saint who died too young. Here is the opening:

You should have seen us when we had our act together, top of our game, toast of the town, walking and talking miracles and—you'd better believe it—the real American thing. We were realer than real, if you ask me, more real than the Original because there were nights back then it felt like he couldn't have opened for us, couldn't have come close to us, not on the best night he ever had, not when you compared it to the nights we were having. I know it sounds crazy but I've got to say it. We went way beyond him. We crossed borders he never traveled, lived in a time he never saw, played in places he couldn't picture.

The nightclub where they perform is in Olongapo, the Philippines, possibly the world's largest brothel town at the time. You see, until 1991 the US Navy had a base at Subic Bay, and there were lots of hormone-fueled young men looking for a good time. The Elvis act acquires a dedicated fan base, extending throughout the country. Locals name their children Elvis. Biggest Elvis tries to improve the lives of the girls who work at the club, and eventually he falls in love. On another level, though, this book is about American foreign policy since World War II. We had the trust and admiration of the world in 1945. But after many good intentions gone awry and support of too many dictators just because they were anti-communist (Trujillo in the Dominican Republic, the Shah of Iran, Marcos in the Philippines, Pinochet in Chile), American foreign policy did as much damage as good in many corners of the planet.

Endnotes

1. Pearl, Nancy, *Book Lust: Recommended Reading for Every Mood, Moment, and Reason*, Seattle, Washington: Sasquatch Books, 2003.
2. Pearl, Nancy, *More Book Lust: 1000 New Reading Recommendations for Every Mood, Moment, and Reason*, Seattle, Washington: Sasquatch Books, 2005.
3. See also: http://americanbookreview.org/100bestlines.asp (accessed March 24, 2013).
4. See also http://fer3.com/arc/m2.aspx/Captain-Ahab-his-Quadrant-Taylor-feb-1998-w1434 (accessed March 13, 2013).

15

Books with a Moral Angle

Common values are a key component of a marriage or a friendship. You can have a discussion that starts out in neutral territory much more easily if one person asks, "What are our common values?" Rather than, "What are the rules?"

But what if we aren't particularly religious? If we don't rely on the Bible, the Torah, the Koran, or the Eightfold Path from Buddhism, where do our values come from? I would assert that in the United States many people who aren't religious still get their moral framework from the Judeo-Christian heritage. This can be supplemented with ideas from many cultures or philosophies. Existentialism: make a choice and take responsibility for your choice. Buddhism: desire is natural. So when you're hungry, eat; when you're tired, sleep; if you want to be loved, love someone. But suffering comes from *selfish* desire. Note the difference.

Setting aside traditional religious works, let us look at a small number of novels that provide us with some moral foundation.

The Dream of Scipio by Iain Pears (2002) has three parallel stories that take place in the same region of southern France. One takes place in the waning days of the Roman Empire (fifth century); the next during the middle of the fourteenth century, when one-third of Europeans died from the plague; and the third during World War II, when northern France was ruled by the Germans and southern France was ruled by the Vichy government under the thumb of the Germans.

In Pears's book the Imperial Roman protagonist ponders our decision-making process when it comes to important things.[1]

> A question of civilized values, he told himself. A question of whether or not one is to take a stand and insist that, despite the times, barbarism must not hold sway. How do we justify calling ourselves civilized, after all? Is it the books we read? The delicacy of our tastes? Our place in continuing a line of belief and of common values that stretch back a thousand years and more? All this, indeed, but what does it mean? How does it show

itself? Are you civilized if you read the right books, yet stand by while your neighbors are massacred, your lands laid waste, your cities brought to ruin?

Do we use the barbarians to control barbarism? Can we exploit them so that they preserve civilized values rather than destroy them? Was the old Athenian right, that taking any side is better than taking no side?

• • • • • • • •

Fyodor Dostoyevsky (1821–1881) and Leo Tolstoy (1828–1910) were the two greatest Russian writers of the nineteenth century. Dostoyevsky's masterpiece was *Prestuplenie i Nakazanie* (*Crime and Punishment*, 1866). Its original title was *The Confession*. It is a detective novel. It is a psychological analysis. It is about crime, punishment, and redemption.

The story deals with a young man named Raskolnikov. Early in the book he kills an old woman who is a pawnbroker. Why? Because he sees her as exploiting the other poor people of his neighborhood. How can he justify this? Part 3, chapter 5, tells us. There is a discussion of an article recently published by Raskolnikov in which he divides people into two groups, the "ordinary" (*obyknovennie*) and "extraordinary" (*neobyknovennie*). The detective investigating the case summarizes: "Ordinary men have to live in submission, have no right to transgress the law, because, don't you see, they are ordinary. But extraordinary men have a right to commit any crime and to transgress the law in any way, just because they are extraordinary."[2,3]

In 1865 and 1866 Napoleon III published a *History of Julius Caesar* in which he declared that Caesar, Charlemagne, and Napoleon Bonaparte charted new paths and played exceptional roles for nations by accelerating events. What did it matter that there were exiles, executions, and *coups d'état*? Dostoyevsky's thinking must have been influenced by this. But Raskolnikov was thinking more of intellectuals, people who shape a nation's *spiritual culture*, such as Kepler, Newton, or Mohammed.[4]

We know who the murderer is from the start, but Dostoyevsky builds suspense in showing how the detective solves the case. Raskolnikov himself provides much of the evidence. And he realizes that no one is above the law. We are all God's creatures. We are all sinners. But we can be redeemed.

Nikolai Nikolaevich Strakhov (1828–1896), a close associate of both Dostoyevsky and Tolstoy, observed that *Crime and Punishment*

> shows us for the first time a nihilist who is unhappy, a nihilist who suffers deeply and humanly.... The author has taken nihilism and carried it to its extreme, to a point from which there is hardly anywhere else to go.... His aim was to show how life and theory struggle within a human soul, to depict that conflict in a situation in which it has reached its highest pitch, and to demonstrate that in the end life wins out. Such was the object of the novel.[5]

• • • • • • • •

The Feast of the Goat by Mario Vargas Llosa (2000) begins: "Urania. Her parents had done her no favor; her name suggested a planet, a mineral, anything but the slender, fine-featured woman with burnished skin and large, dark, rather sad eyes who looked back at her from the mirror. Urania! What an idea for a name." This novel is a very shocking account of life in the Dominican Republic under the dictator Rafael Trujillo, who ruled from 1930 to 1961.

Another novel that describes more horrors is *The Brief Wondrous Life of Oscar Wao* by Junot Diaz (2007). Not only were there informants everywhere and secret police run by the torture hobbyist Johnny Abbes García, but it was understood that all the wives and daughters of government officials were considered available to Trujillo for his sexual urges. To refuse meant that your whole family might be murdered, fed to the sharks, and your house burned down and bulldozed. You and your family could simply be wiped out, as if you had never existed. Even though the reader knows that Trujillo will be assassinated, Vargas Llosa pumps up your adrenaline with his description of how it played out.

A third book about the Trujillo era, *In the Time of the Butterflies* by Julia Alvarez (1994), is also worth reading. It was made into a film starring Salma Hayek in 2001.

Ah, but Trujillo was not all bad, was he? When boatloads of Jewish refugees were refused entry into the United States, Trujillo took some in. What was his motivation? "To lighten the race." He wanted them to intermarry with the Dominicans and produce children with lighter-colored skin.

Amazingly, if you visit the D. R. today you will be impressed with just how friendly the people are. When you buy something in a store, not only do you get your package and your change, but the girl behind the counter will touch you on the forearm and call you "*mi amor.*" It is not a come on. She is just being friendly, Dominican style.

• • • • • • • •

The Great Gatsby by F. Scott Fitzgerald (1925) is a novel about the American dream. Poor boy meets girl, must prove that he is worthy, works hard, then becomes a success. Ah, but at what price? The main character, Jay Gatsby, has become rich as a "businessman," but we are not quite sure if his business dealings were legal. But what does it matter? He has a huge, luxurious estate on Long Island, a beautiful car, and a walk-in closet full of beautiful clothes. This is the Roaring Twenties, an age of fun and optimism, but maybe only if you are rich. Does it matter that his true love, Daisy Buchanan, is married? She is happy to party with Jay Gatsby at his lovely home with his lovely acquired friends.

One undertone of this book is that rich people are different from the rest of us. (See "extraordinary" people in the context of *Crime and Punishment*.) A rich woman might be the actual driver in a hit-and-run accident, but there is no consequence to her, except some stress until the situation gets "fixed." As Fitzgerald says at the end of the novel,

> They were careless people, Tom and Daisy—they smashed up things and creatures and then retreated back into their money or their vast carelessness or whatever it was that kept them together, and let other people clean up the mess they had made.

• • • • • • • •

Heart of Darkness is a short novel by Joseph Conrad, published in 1899. It is one of the most-read works of fiction at the high school and university level. On one hand it is a story about an ivory trader who travels up the Congo River to meet a man named Kurtz who runs a station for acquiring ivory. A variation of the story is presented in Francis Ford Coppola's 1979 film *Apocalypse Now*, which is set in Vietnam and Cambodia.

Key themes of nineteenth- and twentieth-century history are confronted: savagery versus civilization, racism, colonialism, imperialism. "A man's got to do what a man's got to do," versus, "We are our brothers' keepers."

This is worth exploring for the edification of one's mind and one's soul. Good books abound. In the opening pages of *Guns, Germs, and Steel*, Jared Diamond recalls the question posed to him by a man from New Guinea in 1972: " Why is it that you white people developed so much cargo and brought it to New Guinea, but we black people had little cargo of our own?"[6] It took Diamond twenty-five years to come up with a multifaceted answer to the question.

I recommend two other books: 1) *After Tamerlane: the Rise and Fall of Global Empires, 1400–2000* by John Darwin (2008) and 2) *Empire's End: A History of the Far East from High Colonialism to Hong Kong* by John Keay (1997). From a review of Darwin's book:

> It was not Europe's modernity that finally carried the day . . ., "but its superior capacity for organized violence." Our commercial genius and scientific bent certainly counted for something, but not as much as our knack for aggression. By the 19th century, Westerners had figured out how to turn this violence away from themselves—between 1815 and 1914, the continent avoided generalized warfare—and toward the rest of the world.
>
> At this point, the West rises at stunning speed. What still seemed uncertain at the start of the 19th century—the imperial might of the West—had by the end taken on the aura of inevitability. At the same time, missionaries and explorers—some, like David Livingstone, combined the two professions—depicted the newly colonized people in static and primitive terms. In other words, they were peoples who desperately needed to serve a long apprenticeship in Western civilization. . . .
>
> After World War II great imperial powers like England could no longer afford to maintain their far-flung real-estate holdings and their increasingly restless inhabitants. The moral cost had become as intolerable as the financial cost: the war, after all, had been fought on behalf of the cause of freedom. As the rapid process of postwar decolonization made clear, a "European-centered world order was no long sustainable."[7]

• • • • • • • •

A now-famous story of redemption is the story of Oskar Schindler (1908–1974). He is the main character of the novel *Schindler's List* by Thomas Keneally (1982) and the Oscar-winning film starring Liam Neeson and directed by Steven Spielberg (1993). Here was an opportunist, a womanizer, a smooth operator, a "Good Nazi" who accomplished his goal of making a fortune during World War II, but who then plowed his immorally gotten gains into saving twelve hundred Jews from extermination. Read the book, watch the movie again, and keep a box of Kleenex handy.

• • • • • • • •

On the lighter side, another novel with a moral component is *Seeing Calvin Coolidge in a Dream* by John Derbyshire (1996). It is the story of a man who grew up in the People's Republic of China and escaped to Hong Kong (by swimming!) while it was still a British colony. He finds a girlfriend and is working his way up the ladder, starting at a very low rung. His girlfriend decides to marry someone else. Why? Better economic prospects. Many years later, the main character is a successful businessman living in the United States. One of his hobbies, to become "more American," is to visit presidential libraries. He is happily married, as is his long-ago ex-girlfriend who is also living in the United States, but on the opposite coast. But he wants to reconnect with his ex-girlfriend and have an affair. At night he starts having recurrent dreams about President Calvin Coolidge. Silent Cal advises him that the only way to keep anyone from finding out about something you might do that you may not be proud of is—just don't do it.

One of the interesting things about this book is how Derbyshire, a British-born naturalized US citizen, gets into the mind of the Chinese protagonist. It turns out that Derbyshire's wife is Chinese. This same author trained as a mathematician. He has written books such as *Prime Obsession: Bernhard Riemann and the Greatest Unsolved Problem in Mathematics* (2003), which we referenced in chapter 10.

Endnotes

1. Pears, Iain, *The Dream of Scipio*, New York: Riverhead Books, 2002, pp. 171–172.
2. Dostoevsky, Fyodor, *Crime and Punishment*, translated by Constance Garnett, New York: Random House, 1956, p. 234.
3. Dostoevsky, Fëdor Mikhailovich, *Prestuplenie i Nakazanie*, Leningrad: Lenizdat, 1970, p. 252.
4. Grossman, Leonid, *Dostoevsky: a Biography*, translated by Mary Mackler, Indianapolis & New York: Bobbs-Merrill Co., 1975, pp. 359–360.
5. Ibid., p. 355.
6. Diamond, Jared, *Guns, Germs, and Steel: The Fates of Human Societies*, New York and London: W. W. Norton, 1997, p. 14.
7. Zaretsky, Robert, "Is all history the history of empire?" *Houston Chronicle*, April 27, 2008, Zest section, p. 15. Reproduced courtesy of Robert Zaretsky.

16

Lost Books

One time when I was riding the subway in Boston I noticed a woman reading a book with great intensity. It was *The Shadow of the Wind* by Carlos Ruiz Zafón. Finally, she felt me staring at her. She looked up with a somewhat alarmed expression. I smiled slightly, pointed at the book, and said, "Pretty good, isn't it?" When she realized that I was a kindred spirit, not a stalker, she smiled back and said with great conviction, "Yes!"

The book begins:

> I still remember the day my father took me to the Cemetery of Forgotten Books for the first time. It was the early summer of 1945, and we walked through the streets of Barcelona trapped beneath ashen skies as dawn poured over Rambla de Santa Mónica in a wreath of liquid copper.

The narrator and his father stop "in front of a large door of carved wood, blackened by time and humidity. Before us loomed what to my eyes seemed the carcass of a palace, a place of echoes and shadows. . . . A smallish man with vulturine features framed by thick gray hair opened the door." The narrator's father says,

> This is a place of mystery, Daniel, a sanctuary. Every book, every volume you see here, has a soul. The soul of the person who wrote it and of those who read it and lived and dreamed with it. Every time a book changes hands, every time someone runs his eyes down its pages, its spirit grows and strengthens. . . . When a library disappears, or a bookshop closes down, when a book is consigned to oblivion, those of us who know this place, its guardians, make sure that it gets here. In this place, books no longer remembered by anyone, books that are lost in time, live forever, waiting for the day when they will reach a new reader's hands. In the shop we buy and sell them, but in truth books have no owner. Every book you see here has been somebody's best friend. Now they have only us. . . .

La Sombra del Viento or *The Shadow of the Wind* (2004) is the first of three books already published. A fourth may appear. The action takes place from 1917 to 1955, mostly. The other two published volumes are *The Angel's Game* (2010) and *The Prisoner of Heaven* (2012). The order in which you read them does not matter. Some scenes are in Paris, but most of the scenes are set in a foggy dream version or hallucination of Barcelona, where life might be suddenly skewered on an unseen knifeblade or a detective might shoot the man you were talking to yesterday in the face. Zafón creates characters we care about. And we are compelled to join their world. *The Shadow of the Wind* deals primarily with the author of the very rare book that Daniel Sempere chooses to borrow from the Cemetery of Forgotten Books. Why are books by this author so rare? Because someone has been systematically hunting them down and destroying them. What does this person have against these novels or their writer? You will want to find out.

• • • • • • • • •

Poggio Bracciolini (1380–1459) was at one time apostolic secretary to the first Pope John XXIII. This pope was deposed and removed from his position in 1415 by a council that convened in Constance, a German city in the mountains between Switzerland and Germany, on the shores of the Bodensee. The name John XXIII was stricken from the roster of official popes and was not used again until the middle of the twentieth century.

But this is not critical to our narrative. Poggio Bracciolini and other Renaissance-era humanists were most passionate about the recovery of nearly lost documents from ancient times. In January 1417, Poggio visited the Benedictine Abbey of Fulda in central Germany. He discovered the *one* remaining copy of *On the Nature of Things* by Lucretius (ca. 99–55 BC), the Roman poet and philosopher.

An account of this remarkable find and its significance is described by Stephen Greenblatt in his book *The Swerve: How the World Became Modern* (2011). Following Greenblatt's summary, Lucretius's poem introduced Europe to some truly radical thoughts: Everything is made of invisible particles. The elementary particles of matter are eternal. The elementary particles are infinite in number but limited in shape and size. All particles are in motion in an infinite void. Everything comes into being as a result of a "swerve." Lucretius called this the *declinatio, inclinatio,* or *clinamen*. It is a very small deviation from straight-line motion, but the consequence of the collisions of particles brings it about that the "rivers replenish the insatiable sea with plentiful streams of water, that the earth, warmed by the Sun's fostering heat, renews her produce, that the family of animals springs up and thrives, and that the gliding ethereal fires have life."[1] Furthermore, Lucretius asserts that nature ceaselessly experiments. The universe was not created for or about humans. Human society began not in a golden age of tranquility and plenty, but in a primitive battle for survival. The soul dies. There is no afterlife. Death is nothing to us. All organized religions are superstitious delusions. Religions are invariably cruel. There are no angels, demons, or ghosts. The highest goal of human life, therefore, is the enhancement of pleasure and the reduction of pain. The greatest obstacle to pleasure is not pain; it is delusion. Finally, understanding the nature of things generates deep wonder.[2]

How remarkable that Lucretius's nearly lost work was found by a bibliophile who could fully appreciate its significance. Of the books I read in 2012, *The Swerve* was my very favorite.

• • • • • • • •

Word-of-mouth advertising is often the most trustworthy kind of advertising. A recommendation from a friend is worth a lot. A book can be such a friend, recommending books you should get to know. From *The Half-Life of Facts* by Samuel Arbesman I found out about Stuart Kelly's remarkable volume, *The Book of Lost Books*.[3] Sometime prior to turning fourteen, Kelly had become obsessed with reading, or at least finding, all of the books written by a particular author. His passion, or obsession, led him to compile lists of books attributed to certain authors, but were lost, or books planned by certain authors and never written.

There are eighty-one chapters in Kelly's book, most covering one author at a time. Aeschylus (ca. 525–456 BC) wrote eighty plays. Seven have survived. Sophocles (ca. 495–406 BC) wrote 120 plays; seven have survived. Euripides (ca. 480–406 BC) wrote more than ninety; eighteen have survived. Of course there are chapters on Chaucer, Shakespeare, and Milton.

A more obscure writer was Sir Thomas Urquhart (1611–1660), a British royalist who opposed Oliver Cromwell, the military commander in the First English Civil War (1642–1646) and later Lord Protector of the Commonwealth of England, Scotland, and Ireland. As a writer Urquhart was known for hyperbole. Upon the restoration of King Charles II in 1660, Urquhart died "in a fit of excessive laughter."[4]

Racine, Leibniz, Goethe, Jane Austen—all were well-known writers with works that were lost or that failed to reach fruition.

Nikolai Gogol (1809–1852) was the author of one of the greatest Russian novels of the nineteenth century, *Dead Souls* (1842). It deals with a wheeler-dealer named Chichikov who wishes to acquire a reputation as an important person. That would normally require that he be a major landowner, which means he would be the owner of all the *serfs* that work those lands. (The serfs were not emancipated until 1861.) Chichikov does not really want the *land*. He wants the reputation. So after getting to know the landowners near a particular town, he offers to buy the serfs that have died, but whose names are still part of the census. Everybody wins. The government still gets its tax, the landowners pay less tax, and Chichikov becomes somewhat of a big shot.

Gogol worked on Part II of *Dead Souls* for several years. But then he came to believe, under the influence of a priest named Father Matthew Konstantinovky, that everything except the Russian Orthodox Church was inspired by Satan. As penance Gogol fed the manuscript pages of the unpublished second part of *Dead Souls* to the fireplace. Then, after nine days of self-enforced starvation, he died.[5]

Here is the opening of the chapter on Sir Richard Burton (1821–1890), whom we met in chapter 4 of this book: "Witnessing cannibalism was one of the few ambitions Sir Richard Burton failed to satisfy."[6] He mastered many languages, translated the *Kama Sutra* and *One Thousand and One Nights*, was the British consul in Damascus and other places, and somehow managed to find a wife who was an ideal companion. However, upon his death, she destroyed

Burton's journals from 1872 to 1890, their letters, his unfinished manuscripts, lists of Syrian proverbs, notes on the eunuch trade, translations of ancient Roman and Renaissance poets, and studies of polygamy.[7]

The final chapter in Kelly's book deals with George Perec (1936–1982), a French writer and filmmaker. He was a member of OuLiPo, l'Ouvroir de Littérature Potentielle, the Workshop of Potential Literature. His novel *La Disparition* (1969) is written without ever using the letter *e*. It is a thriller about the missing Anton Vowl.[8] Another member of OuLiPo was Italo Calvino, to whom we turn in the next chapter.

Endnotes

1. Greenblatt, Stephen, *The Swerve: How the World Became Modern*, New York and London: W. W. Norton, 2011, pp. 188–189.
2. Ibid., pp. 185–199.
3. Kelly, Stuart, *The Book of Lost Books: An Incomplete History of All the Great Books You'll Never Read*, New York: Random House, 2005.
4. Ibid., p. 166.
5. Ibid., p. 259.
6. Ibid., p. 280.
7. Ibid., pp. 280–284.
8. Ibid., p. 333.

17
A Novel Concept

To get lost in a book is to have an adventure without ever leaving the house. Usually, it is just because you like the story, but sometimes it is because you like the language, style, or sense of humor of the author. Sometimes it is because you appreciate a challenge.

One of my favorite authors is Italo Calvino (1923–1985). His collection of science-fiction stories *Cosmicomics* was the first of half a dozen of his books I have read. The story about the last dinosaur is not really about dinosaurs. It is about racism. The story "The Light-Years" is not really about the expansion of the universe. It is about being overly concerned about other people's opinions of you and how nearly impossible it is to fully explain yourself.

In 1981 I first read the English translation of his most clever book. It begins:

> You are about to begin reading Italo Calvino's new novel, *If on a winter's night a traveler*. Relax. Concentrate. Dispel every other thought. Let the world around you fade. Best to close the door; the TV is always on in the next room. Tell the others right away, "No, I don't want to watch TV!" Raise your voice—they won't hear you otherwise—"I'm reading! I don't want to be disturbed!"

Each of the first ten chapters consists of two parts. Each first part is about two people who are looking for a good book to read. Each second part is the first chapter of a different story. Some of the stories have certain similarities. There is also an international translation conspiracy to unravel. Finally, at the end of the book, it turns out that *you* are one of the two readers looking for a good book to read.

Calvino was a member of OuLiPo, the Workshop of Potential Literature. He endeavored to create labyrinths in which the reader wanted to get lost.

• • • • • • • •

17 A Novel Concept

One of the most famous novels of the twentieth century was *Ulysses*, by the Irish writer James Joyce (1882–1941). It takes place in a twenty-four-hour period beginning June 16, 1904. Why this date? Joyce was honoring the date he and Nora Barnacle, his future wife and mother of his two children, had their first date. The main character of the book is one Leopold Bloom, and in honor of this character June 16 is known today in literary circles as Bloomsday.

The prototype for Bloom is Homer's hero Ulysses. But instead of wandering the world for twenty years, Bloom wanders around Dublin trying to drum up advertising revenue for the newspaper where he works. ("What is home without Plumtree's Potted Meat? Incomplete. With it an abode of bliss.")

The prototype of Ulysses's son Telemachus is Stephen Dedalus, who, at the beginning of the novel is affected by the "Agenbite of Inwit" (remorse of conscience); on his mother's deathbed he refused to pray for her. Few of us would know that *Agenbite of Inwit* is a prose work from ca. 1340 written in the Kentish dialect of Middle English. This is one of hundreds of literary references in this 768-page novel.

Ulysses is written in eighteen chapters, each in its own style. Chapter 7 (Aeolus) is delineated with newspaper headlines. There is a monster every third chapter. Molly is Bloom's wife, the analog of Penelope (the wife of Ulysses). One big difference is that Penelope was faithful to Ulysses. Molly has been having an affair with a retired boxer named Blazes Boylan. Bloom is aware of this, and much to his dismay he keeps hearing references to Boylan and bumping into him all day.

According to Joyce, his wife spoke in a particular rambling style. This is reproduced in the final chapter of *Ulysses*, with many run-on sentences devoid of punctuation, except for one period at the very end. It is a soliloquy given by Molly Bloom, in bed, with Leopold Bloom sleeping next to her (but with his head at the foot of the bed). She reminisces about how she and Bloom fell in love long ago. The chapter ends with the most positive word in the English language:

> . . . I was a Flower of the mountain yes when I put the rose in my hair like the Andalusian girls used or shall I wear a red yes and how he kissed me under the Moorish wall and I thought well as well him as another and then I asked him with my eyes to ask again yes and then he asked me would I yes to say yes my mountain flower and first I put my arms around him yes and drew him down to me so he could feel my breasts all perfume yes and his heart was going like mad and yes I said yes I will Yes.

Ulysses was published on February 2, 1922, Joyce's fortieth birthday. The publisher was Sylvia Beach, the proprietor of a Paris bookstore called Shakespeare and Company. It was a place many artists picked up their mail, and in that way the members of the "Lost Generation" met each other and encouraged each other in their endeavors. Amongst the younger members of this circle were Ernest Hemingway and F. Scott Fitzgerald. Elder members were Gertrude Stein and James Joyce. I highly recommend a monograph by Noel Riley Fitch called *Sylvia Beach and the Lost Generation*.[1] Woody Allen's more recent film *Midnight in Paris* (2011) covers some of the same ground.

Ulysses is a difficult novel to read, primarily because of its stream-of-consciousness style. I did not start getting the hang of it until the third time I read it, and I have read it twice more since then. A good book *about* the book is by Joyce's friend Frank Budgen: *James Joyce and the Making of Ulysses and other Writings*.[2] Since Budgen was a personal friend of Joyce, there are many interesting firsthand anecdotes. For example, one day in Zurich the two bumped into each other and Budgen asked Joyce how the book was going. Joyce said that he had accomplished a great deal that day. He had almost written two sentences. Why so little? He had the words but was looking for the right order.[3] What are the two sentences, you are wondering? "Perfume of embraces all him assailed. With hungered flesh obscurely, he mutely craved to adore."[4]

Ulysses is a book one can spend years studying. Joyce would be pleased with such dedication. He once said, "The demand that I make of my reader is that he should devote his whole Life to reading my works."[5] Someday, before I get much older, I want to take a trip to Dublin, overlapping Bloomsday, to take a literary tour of the city and follow in the footsteps of Leopold Bloom.

• • • • • • • •

I learned about Alexander Theroux's 1981 novel *Darconville's Cat* from Nancy Pearl's first book of reading selections, *Book Lust*. This one proved to be quite a challenge, as she warned. My abridged dictionary was not up to the task of providing definitions of many words. In the opening fifty pages we come across ipsedixitist, pneumatomachian, cataphatic, guerdon, tracassaries, manciple, glyptodont, protrepticos, senseflectum, jupon, sphragistic, quidnunc, and glossoepiglottic, none of which I knew, and less than half of which are in my abridged dictionary. Many words in *Darconville's Cat* are either made up from Greek parts or are genuine words that are archaic or just generally unused by normal people, such as neautontimoroumenotic (p. 239).

This is a book nominally about Alaric Darconville, a young literature professor at an all-girls college in Virginia. But with all the stream of consciousness, the lists of people and books, and plays on words in multiple languages, it is really something else. It reminded me of *Ulysses*. There is even reference to *Agenbite of Inwit* (p. 443).

On p. 303 Theroux writes, "The Virginians in this particular area, briefly, had a marvelous idea of themselves. What was past was perfect." And I see I have written in the margin, "His past imperfect made his present tense." Not quite sure why.

The most curious thing occurs about two-thirds of the way through the book. One of the key parts of the book is Darconville's romance with one of his students. When he finds out that she is breaking up with him and getting together with a guy from her hometown, about whom she said almost nothing good, he is understandably fuming. The reader then turns the page and is confronted with two pages that are completely and totally black! (See Figure 17.1.) In chapter 93 a misogynistic professor from Harvard gives Darconville many suggestions of what to do in response, including dropping raccoons full of diphtheria viruses down her chimney. Why is this?

17 A Novel Concept

Figure 17.1 Pages 482–483 of *Darconville's Cat* by Alexander Theroux. This immediately follows after the protagonist learns the identity of the man his girlfriend has dumped him for.

To quote Sherlock Holmes (p. 618), "There are certain crimes which the law cannot touch and which therefore, to some extent, justify private revenge."

• • • • • • • • •

Leonid Tsypkin (1926–1982) was a Soviet medical researcher who wrote prose in his spare time. *Summer in Baden-Baden* was written between 1977 and 1980. A copy was smuggled out of the Soviet Union in 1981 and it began to be serialized in a Russian weekly in New York, *Novaya Gazeta*. Tsypkin died of a heart attack a week after the first installment appeared.

The novel is short, only 146 pages, with no chapter divisions. It has a double narrative. One is about Dostoyevsky and his second wife, who are on their way to Germany in 1867, where they will spend four years. The other narrative is about a dedicated fan of Dostoyevsky who takes a train from Moscow to the once and future St. Petersburg, where Dostoyevsky principally lived and where he died in 1881. A common theme of the second narrative deals with Dostoyevsky's anti-Semitism and the irony that he had many Jewish fans, including Tsypkin himself and the well-known Dostoyevsky biographer Leonid Grossman (1888–1965).[6]

The novel is written in a style I have never seen before. Complete with punctuation (unlike the final chapter of *Ulysses*), every paragraph is one sentence, sometimes one very long sentence. I could only read it ten pages at a time. Here is part of the opening paragraph:

> I was on a train, travelling by day, but it was winter-time—late December, the very depths—and to add to it the train was heading north—to Leningrad—so it was quickly

darkening on the other side of the windows—bright lights of Moscow stations flashing into view and vanishing again behind me like the scattering of some invisible hand—each snow-veiled suburban platform with its fleeting row of lamps melting into one fiery ribbon—the dull drone of a station rushing past, as if the train were roaring over a bridge—the sound muffled by the double-glazed windows with frames not quite hermetically sealed into fogged-up, half-frozen panes of glass—pierced even so by the station-lights forcefully etching their line of fire—and beyond, the sense of boundless snowy wastes—and the violent sway of the carriage from side to side—pitching and rolling—especially in the end corridor—and outside, once complete darkness had fallen and only the hazy whiteness of snow was visible and the suburban dachas had come to an end and in the window along with me was the reflection of the carriage with its ceiling-lights and seated passengers, I took from the suitcase in the rack above me a book I had already started to read in Moscow and which I had brought especially for the journey to Leningrad....

• • • • • • • •

William Faulkner (1895–1962) was the 1949 winner of the Nobel Prize in Literature. His most famous novel is perhaps *The Sound and the Fury* (1929), which is set in a fictional place in Mississippi called Yoknapatawpha County. Thanks to its stream-of-consciousness style, the use of multiple unreliable narrators, the tendency to jump back and forth in time, and Faulker's unique way of describing even simple things, it is not an easy read.

An annual contest to write like Faulkner began in 1990. The 1998 winner, Robert Blake, who normally writes for medical journals, managed to use some standard Faulkner buzzwords, such as "abnegation," "abrogation," and "viscera," but could not use "effluvium," a Faulkner favorite. Here is Blake's entry, which is called "Pile On," in reference to Faulkner's *Pylon* (1935):

Knowing knows before hearing hears, recollection exudes from the congealed entanglement, esasculate in the indomitable odor of mansweat; remembering before knowing: hands splayed on bended knees, semi-crouched in rapt immobility, forward-leaning into the ponderous nocturnal autumn air, in furious anticipation of arrested inertia, incipient savagery, luminous in the brooding dusk-dark; forwardmoving preemptorily with the sound, an inviolate sonorous command, refusing abnegation, compelling allegiance, doomed in the primordial obdurate masculinity; receiving the thrusted leather oblong not-trophy, neither chalice, but rather palpable symbol of insatiable honor, impregnable, invincible but ephemeral; viscera thrusted, arms engulfing as a lover's embrace, but futile; forwardmoving with escalating fury inexorably toward the armor-clad foe, non-apparitional, voracious, implacable, intractable, incorrigible and girded for the assault in resplendent triumph; arrested in stark, abrupt and utter abrogation of motion, profound dissolution, sudden and complete; and now cohered with the hard, immutable earth; with the penetrant whistling infiltrating through the laboriously unlimbering extrication of virile man-flesh to the abject fury of disembodied surrender; and then, with resolute, authoritative finality, the hearing: "second down."[7]

· · · · · · · · ·

By contrast, Ernest Hemingway is alleged to have taken up a challenge to write the shortest possible story, coming up with these six words: "For sale, baby shoes, never worn." But this anecdote is probably not true.[8]

Endnotes

1. Fitch, Noel Riley, *Sylvia Beach and the Lost Generation: A History of Literary Paris in the Twenties and Thirties*, New York and London: W. W. Norton, 1983.
2. Budgen, Frank, *James Joyce and the Making of 'Ulysses' and other writings*, Oxford and New York: Oxford University Press, 1989.
3. Ibid., p. 20.
4. This is found on p. 166 of my Random House edition of *Ulysses*.
5. http://www.brainyquote.com/quotes/authors/j/james_joyce_2.html (accessed May 3, 2013).
6. Tsypkin, Leonid, *Summer in Baden-Baden: A Novel*, translated by Roger and Angela Keys, New York: New Directions, 2001.
7. Holland, Gina, "Physician pens winning entry in Faulkner contest," *The Seattle Times*, July 26, 1998, p. A5. Reproduced courtesy of Robert L. Blake, Jr., and Yoknapatawpha Press.
8. Haglund, David, "Did Hemingway really write his famous six-word story?" *Slate*, posted January 31, 2013.

18

Recommended Nonfiction

Asking Some of the Right Questions

There are thousands of books worth reading. I can only recommend *some* that I found interesting. Let us start with a new category of book that is a combination of applied statistics and psychology. How should we analyze certain aspects of human behavior and human institutions? A main problem is that people behave differently if they know they are being observed, so how to find out their true tendencies?

Steven Levitt and Stephen Dubner hit home runs with their two books *Freakonomics* and *Super Freakonomics*.[1,2] As discussed already in chapter 13, these are the guys who analyzed test results in the Chicago Public School system from the 1990s and discovered that some teachers were cheating for their students. In spite of the publicity for this cheating and the consequences (people getting fired!), some teachers in Atlanta schools were found doing exactly the same thing in 2013.

Levitt and Dubner also discovered one significant cause of the drop in crime during the 1990s. It was *not* due to what you might think: more police, stricter gun control, increased punishment. It was attributable to access to abortion facilities starting in the 1970s. With fewer *unwanted* children (who grow up unloved and unsupervised) there are fewer future criminals.

Malcolm Gladwell is a writer for the *New Yorker* who has the ability to ask large questions and to deliver a narrative in nice, bite-size pieces for our consumption. And I swear, this guy could write an article about paint drying and make it interesting. He burst onto the national scene with his book *The Tipping Point*.[3] One of the many topics covered in this book (which reputedly garnered the author a million-dollar advance) is about graffiti on metro systems and the connection to street crime. In New York City it was decided that if a subway car came back from a day's travel with graffiti on it, it would not go back out until the graffiti was eliminated. The graffiti was symbolic of the collapse of the system and visible to all (p. 142). Next came stomping out fare-beating. It turned out that one in seven fare-beaters had an outstanding warrant for a previous crime, so it was cost-effective to arrest the fare-beaters because many serious criminals were removed from the streets (p. 145).

How many Facebook friends do you have? Seven hundred? I knew a world-class track star with over two thousand. At the time of this writing the allowed limit is five thousand. But Dunbar's Rule (p. 179 of *The Tipping Point*) suggests that you cannot carry on more than 150 relationships of all types. This holds whether you are a modern city dweller or a villager in rural Australia, New Guinea, Greenland, or Tierra del Fuego.

In his second best seller, *Blink*,[4] Gladwell explains "the Warren Harding Effect," namely, we have this innate prejudice that *tall*, good-looking men are natural born leaders. As a state senator and US senator from Ohio, Harding did all right, but by the time he was president and surrounded himself with corrupt cronies, he was out of his league.

In chapter 1 of *Outliers*, Gladwell explains the Matthew Effect, that the "haves" keep getting opportunities that the "have nots" do not get.[5] In sports such as baseball, soccer, and hockey, if the cutoff date for being included on a team is January 1, those children born in the early part of the year are just enough older to be faster and more coordinated than their friends who were born in the second half of the year. These young athletes are more likely to attract the positive attention of coaches and get selected for all-star teams. This holds all the way to the professional ranks.

In chapter 2 of *Outliers*, Gladwell discusses the Ten Thousand Hour Rule. The implication is that in music, sports, and many other activities, you cannot reach your peak until you have put in ten thousand hours of practice or performance. Four hours a day, five days a week for a year is about one thousand hours. After ten years you have put in your ten thousand hours. While this might have worked well for the Beatles and Tiger Woods, hours alone with the best teachers and coaches would not be enough for the majority of us to be The Best, though we certainly may have reached *our* personal potential.

I have only one friend who grew up in Transylvania, and he is one of the leaders in the new discipline of network science. László Barabási and I were faculty in the physics department at the University of Notre Dame. Now he is a professor at Northeastern University in Boston. In his 2003 book *Linked*, he describes how everything is connected to everything else.[6] For example, the famous Hungarian mathematician Paul Erdös (1913–1996) published over fifteen hundred papers with 507 coauthors.[7] To have an "Erdös Number" of one meant that you published a paper with Erdös himself. If you have an Erdös Number of two, you published a paper with someone who published a paper with him. It is a mark of pride of mathematicians to have a low Erdös Number.

You may have heard of the game Six Degrees of Kevin Bacon. The idea is to figure out how closely or remotely connected another actor is from Kevin Bacon, using movie cast lists. A website now allows you to do this conveniently for any pair of people in movies or television, *but* of course it is only as accurate as the information in the Internet Movie Database (imdb.com).[8] For example, I was in episode one of the 1991 PBS series *The Astronomers*, narrated by Richard Chamberlain.

According to the Oracle of Bacon website (using movies alone) I have a "Richard Chamberlain Number" of three, which is obtained from my having been an extra in a Chilean film called *Machuca* (2004). This involves intermediate links to actors Federico Luppi and Eusebio Lázaro. But if we include television shows, I have a Richard Chamberlain Number of one.

In *Connected*, Nicholas Christakis and James Fowler look at many aspects of social networks.[9] A particularly interesting section (pp. 86–91) is about the life expectancies of married people versus widows and widowers. In short, marriage adds seven years to a man's life and two years to a woman's life. A woman whose husband dies does not have a significantly higher chance of dying in the first couple years of widowhood, but widowers die in much greater numbers shortly after becoming widowers. But while white couples suffer a widowhood effect, black couples do not. In interracial couples men married to black women do not experience a widowhood effect, but men married to white women did. This could be due to different social structure in black families. Families of black wives are generally more supportive during the bereavement of husbands than are families of white wives.

> We do not know yet about same-sex marriage. It could be that married homosexual men each gain seven years of life, and homosexual women each gain two years, just like heterosexual men and women. But it is also possible that married homosexual men gain two years, while homosexual women gain seven. If this were the case, it would mean that it is not marriage per se that is salubrious but, rather, marriage to a woman.[10]

One of my favorite books of 2012 was *The Signal and the Noise* by statistician Nate Silver.[11] He is the blogger for the *New York Times* who correctly predicted forty-nine state-by-state winners of the 2008 and 2012 presidential elections. In 2012 he could not make a prediction for Florida because statistically it was a dead heat.

People might think that Silver is a diehard Democrat, and maybe he is, since Democrats took solace in his daily updates during the 2012 election season. But his book is devoid of any political axe grinding. His point of view is simple—never mind what you *think* is the case. You have to let the data speak for themselves. His book deals with making money playing online poker, predictions for earthquakes, climate change, the stock market, baseball statistics that count (see the 2011 movie *Moneyball*), and even testing whether television "experts" are better predictors than just flipping a coin. Read this book and others listed above. Then tilt your head slightly sideways and ask yourself what kinds of questions and analysis will help us make more sense of the world, since the world is run by *people*.

The Moral Animal

One day in 2003 my friend Tim Abbott loaned me his copy of Robert Wright's book *The Moral Animal*.[12] I found so much of it thought-provoking that I could not help underlining a lot of passages. I bought Tim a pristine new copy. But he wanted to see what I had underlined in his

Figure 18.1 Hugo Schwarz (1953–2006), astronomer and avid reader.

old copy. After that the original copy was mine. I loaned it to our friend Hugo Schwarz, who made his own commentary throughout. Then it got loaned to Hektor Monteiro, then to my elder brother, who in turn loaned it to a friend of his. And finally I got it back.

What was it about this book that we all found it so interesting? It is about *why we are the way we are*. In addition to much discussion of Darwinism, there is also a lot of background on Darwin himself and his relationship with his wife Emma, who was also his first cousin.

Here is a sampling of tidbits and quotes. In a study of thirty-seven cultures around the world, evolutionary psychologist David Buss found that in every culture females placed more emphasis than males on a potential mate's *financial* prospects (p. 60). The number of sperm cells per cubic centimeter in a man's ejaculate does not primarily depend on how long it has been since he last had sex. It "depends heavily on the amount of time a man's mate has been out of his sight lately" (p. 71). Is it "natural" for men to be monogamous? "A huge majority—980 of the 1,154 past or present societies for which anthropologists have data—have permitted a man to have more than one wife" (p. 90). What does Wright think about a lot of feminist scholarship? "Feminists have written articles and books denouncing 'biological determinism' without bothering to understand biology or determinism" (p. 137).

Wright argues (p. 280) that

> the human brain is, in large part, a machine for winning arguments, a machine for convincing others that its owner is in the right—and thus a machine for convincing its owner of the same thing. The brain is like a good lawyer: given any set of interests to defend, it sets about convincing the world of their moral and logical worth, regardless of whether they in fact have any of either. Like a lawyer, the human brain wants victory, not truth; and, like a lawyer, it is sometimes more admirable for skill than for virtue.

Some more quotes: "Brotherly love tends to make society fall apart.... And in a society where no one gets punished for anything, repugnant behavior will grow" (p. 346). "Sensual pleasures are the whip natural selection uses to control us, to keep us in the thrall of its warped values system" (p. 370). At the end of the book (p. 377), Wright asserts that to be a truly moral animal one must, "Go above and beyond the call of a smoothly functioning conscience; help those who aren't likely to help you in return, and do so when nobody's watching."

The Wealth and Poverty of Nations

The best-read person I have known was Hugo Schwarz.[13] His native language was Dutch, and he was proud to assert that it was the best language in the world for swearing. I once asked him how many books he had read. He said it was hard to estimate, but the answer was somewhere between three thousand and four thousand. If a book was written in Dutch, English, German, or Spanish, he preferred to read it in the original.

One of Hugo's excellent suggestions was *The Wealth and Poverty of Nations* by Harvard professor emeritus of history and economics David Landes.[14] Have you ever wondered why some nations are so rich and others perpetually poor? Landes authoritatively addresses the answers as he sees them.

In *Guns, Germs, and Steel* a man from New Guinea asked Jared Diamond why white people had sailed around the world for trade, while black people had not. Landes suggests that one of the reasons was that Europeans were passionately *curious* about other peoples and societies.[15] The Chinese, on the other hand, had a very strong sense of superiority and consequently no curiosity about others. I remember in my year-long college class on Chinese history the professor noted that Queen Victoria was regarded as a "nice barbarian lady." It was just their way of thinking.

When British engineers were building the Indian railways, they acknowledged that the laborers would move earth and rock by hand. The engineers took it for granted that they would use wheelbarrows. But Indians were used to carrying baskets on their heads and there was even one report of Indian laborers placing barrows on their head rather than wheel them![16]

What is a mark of a powerful autocrat? Consider the construction of the first railroad line in Russia, which was to run from Moscow to St. Petersburg. "The tsar was asked to select the route. He took a ruler and drew a straight line between the two cities. But the tip of one finger stuck out, so the line was built with one curved section."[17]

Did you know that in colonial Buenos Aires a horseshoe cost several times the price of a horse?[18] Did you know that immigration to the United States peaked in the first decade of the twentieth century (879.5 thousand)?[19] Did you know that Saudi Arabia, despite having a lot of nearby sand, imported a great deal of beach sand from Australia?[20] Hugo, who had a house on La Palma in the Canary Islands, told me that on Tenerife they imported Saharan sand from the African mainland, but they inadvertently imported a lot of Saharan scorpions!

As I was nearing the end of Landes's book I came into Hugo's office one day and said, "I just read this great quote in a footnote." And he knew which quote I meant even before I elaborated any further. It was by a Harumpher of the First Class, Sir George Robinson, superintendent of British traders at Canton, who groveled thusly to Lord Palmerston in London:

> I trust it is not necessary for me to add anything like an assurance of the most profound deference and respect with which I shall implicitly obey and execute the very spirit of such instructions as I may have the honour to receive, on this or any other point. Strict undeviating obedience to the orders and directions of which I may be in possession ... is the foundation on which I build. ....[21]

Can't We All Get Along?

The *Ornament of the World* was the description of Cordoba by the tenth-century Saxon writer Hroswitha. In her book of the same name, María Rosa Menocal masterfully relates the history of Spain from the seventh through the fifteenth centuries.[22] Set aside what you thought you knew about the Dark Ages. Whereas in Christian Europe the largest library at that time had four *hundred* volumes, in tenth-century Cordoba there were seventy libraries. The caliphal library had some *four hundred thousand* volumes.

It was through Spain that works from ancient Greece made their appearance in Europe. Jews, Christians, and Muslims worked together on scholarly and political matters. The caliph Abd al-Rahman III (who ruled Cordoba from 912 to 961) appointed Hasdai, the son of Isaac, the prince of the Jewish community and his foreign secretary. "Hasdai spoke and wrote with elegance and subtlety ... and possessed a profound knowledge of everything in Islamic and Andalusian culture and politics that a caliph needed in his public transactions."[23] Samuel the Nagid (Samuel ibn Nagrila), born in 993, appointed the head of the Jewish community of Granada, was also a noted poet. "For the first time in a thousand years, Hebrew was brought out of the confines of the synagogue and made as versatile as the Arabic that was the native language of the Andalusian Jewish community."[24] Toledo became the European capital of translations. The Christian king of Castile, Alfonso the Wise (mid-thirteenth century), was a generous patron of astronomy.

Eventually, however, Christians and fundamentalist Muslims from North Africa put an end to the culture of tolerance in medieval Spain. In 1492 Granada ended its reign as the last Muslim capital of Europe.

Fast forward four hundred years. In October of 1905 a boy named Lev Nussimbaum was born to the wife of a wealthy oilman and grew up in Baku, Azerbaijan. As an adult, Lev fashioned himself as a Muslim prince from Central Asia. He was known to the literary world as Kurban Said but preferred the name Mohammad Essad Bey. This chameleon of a person, though Jewish, flourished in Berlin during the 1920s and 1930s. One of his favorite places to hang out was referred to as Café Megalomania. "Hide in plain sight" was apparently his motto. Nussimbaum died in Positano, Italy, in 1942. He is most famous today for having written the "national novel of Georgia" (the country, not the American state). This novel, *Ali and Nino*, is about the love affair of an Azerbaijani Muslim boy and a Christian girl from Georgia. It was his plea for tolerance in a very intolerant world. I highly recommend Tom Reiss's biography *The Orientalist*.[25] As one friend of mine said after I loaned him my copy, "This is the book I've been waiting to read my whole life."

Final Recommendations

After his best seller *Guns, Germs, and Steel*, Jared Diamond wrote the equally engrossing book *Collapse*.[26] He writes about many societies that did not survive: the Norse settlements on Greenland, Easter Island, the Maya, the genocide in Rwanda, to name a few. Of the societies that have survived, one of the most curious is the tiny, isolated Southwest Pacific island of Tikopia.[27] It has an area of just 1.8 square miles and has supported a population of twelve hundred for about three thousand years! They solved two problems: sufficient food supply and population control. As you can imagine, fruits and vegetables are grown everywhere they can be grown. They even have multistory orchards. Permission to catch or eat fish prevented overfishing. About four hundred years ago all the pigs on the island were killed because it turns out that it takes about ten pounds of vegetables edible to humans to produce one pound of pork. The pigs competed too much with humans for sources of food. Instead, the people eat fish, shellfish, and turtles. There have been seven identified methods of population control, including abortion, infanticide, and suicide. More common than explicit suicide was "virtual suicide" by setting out on dangerous overseas voyages. Though some Tikopians now live in the less densely populated Solomon Islands, the chiefs limit the number of people living on Tikopia to 1,115.

Having lived in Chile for more than four years, I was simultaneously fascinated and righteously indignant to learn what I learned from Peter Kornbluh's book *The Pinochet File*.[28] September 11 is not just a day of infamy in the United States. On September 11, 1973, Chilean president Salvador Allende was overthrown in a coup led by Augusto Pinochet. To some, the acts committed by Pinochet and his associates saved the country from the menace of Communism. But over three thousand people just disappeared without due process of law.

During the Clinton administration many documents from that era were declassified. They are still redacted, but one can see to what extent the United States encouraged the eventual overthrow of Allende, starting from discussions in the Oval Office between Henry Kissinger and President Richard Nixon. Even after Allende was dead, his former associates were hunted down and murdered as part of Operation Condor, including the 1976 assassination in Washington, DC, of Orlando Letelier, the former Chilean ambassador to the United States.

I highly recommend this book. Let me make only one specific reference. On pages 68–71 we see a redacted secret CIA cable from Santiago. The date is October 19, 1970, a month and a half after the election that made Allende president. This cable deals with the kidnapping of a general and the initiation of a military coup. We see reference to General Schaffhauser, the chief of staff of the army. A good friend of mine and one of my closest scientific colleagues is married to a

Chilean woman whose family name is Schaffhauser. One Thanksgiving Day in La Serena, Chile, at the house of another astronomer, my friend's wife asserted that nothing really bad happened when Pinochet was the boss. Well, gee, nothing bad happened in *her* family because they had connections at the highest level!

If you have been intrigued by some of the quotes from books discussed above, perhaps you will simply trust me that these are also worth reading:

- *The Song Lines* by Bruce Chatwin, New York: Penguin, 1987. A book about nomads, particularly the Aborigines of Australia. As a book review from the *Boston Globe* notes, it is a blend of travelogue, memoir, history, philosophy, and science.
- *The Making of the Atomic Bomb* by Richard Rhodes, New York: Simon & Schuster, 1986. One peripheral question to ask is why there were so many Hungarians who worked on the Manhattan Project, and why, if someone's first language is Hungarian, he or she can never speak English without some level of Hungarian accent. A related book is *The Girls of Atomic City: The Untold Story of the Women Who Helped Win World War II* by Denise Kiernan, New York, Simon & Schuster, 2013. This is an account of the secret lab in Oak Ridge, Tennessee, that made fissionable material for the Manhattan Project—so secret that many of the people who worked there did not know what the ultimate purpose of their work was!
- *Lenin's Tomb: The Last Days of the Soviet Empire* by David Remnick, New York: Random House, 1993. In a world increasingly filled with suicide bombers, green-on-green attacks in Afghanistan, and an Arab Spring failing to bring peace and stability, we may long for a time when we were the Good Guys and the Soviets were our respected adversaries. This is a brilliant account of the end of the Soviet Union.

Endnotes

1. Levitt, Steven D. & Stephen J. Dubner, *Freakonomics: A Rogue Economist Explores the Hidden Side of Everything*, New York: Harper Collins, 2005.
2. Levitt, Steven D. & Stephen J. Dubner, *Super Freakonomics: Global Cooling, Patriotic Prostitutes, and Why Suicide Bombers Should Buy Life Insurance*, New York: Harper Collins, 2009.
3. Gladwell, Malcolm, *The Tipping Point: How Little Things Can Make a Big Difference*, New York, Boston: Little, Brown and Co., 2000.
4. Gladwell, Malcolm, *Blink: The Power of Thinking without Thinking*, New York, Boston: Little, Brown and Co., 2005.
5. Gladwell, Malcolm, *Outliers: The Story of Success*, New York, Boston, London: Little, Brown and Co., 2008.
6. Barabási, Albert-László, *Linked: How Everything Is Connected to Everything Else and What It Means for Business, Science, and Everyday Life*, New York: Plume, 2003.
7. Hoffman, Paul, *The Man Who Loved Only Numbers: The Story of Paul Erdös and the Search for Mathematical Truth*, 1998.

8. http://oracleofbacon.org/
9. Christakis, Nicholas A. & James H. Fowler, *Connected: The Surprising Power of Our Social Networks and How They Shape Our Lives*, New York, Boston, London: Little, Brown and Co., 2009.
10. Ibid., p. 89.
11. Silver, Nate, *The Signal and the Noise: Why So Many Predictions Fail—but Some Don't*, New York: Penguin Press, 2012.
12. Wright, Robert, *The Moral Animal: Evolutionary Psychology and Everyday Life*, New York: Vintage, 1994.
13. Krisciunas, Kevin, "Hugo Emond Schwarz (1953–2006)," *Bulletin of the American Astron. Society*, 39, no. 4 (December 2007), pp. 1079–1081.
14. Landes, David S., *The Wealth and Poverty of Nations: Why Some Are So Rich and Some So Poor*, London: Abacus, 1998.
15. Ibid., p. 164.
16. Ibid., p. 229.
17. Ibid., p. 265.
18. Ibid., p. 316.
19. Ibid., p. 321.
20. Ibid., p. 408.
21. Ibid., pp. 426–427.
22. Menocal, María Rosa, *The Ornament of the World: How Muslims, Jews and Christians Created a Culture of Tolerance in Medieval Spain*, New York, Boston, London: Back Bay Books, 2002.
23. Ibid., p. 80.
24. Ibid., p. 109.
25. Reiss, Tom, *The Orientalist: Solving the Mystery of a Strange and Dangerous Life*, New York: Random House, 2006.
26. Diamond, Jared, *Collapse: How Societies choose to Fail or Succeed*, New York: Penguin, 2011.
27. Ibid., pp. 286–293.
28. Kornbluh, Peter, *The Pinochet File: A Declassified Dossier on Atrocity and Accountability*, New York and London: The New Press, 2004.

Epilogue: The Importance of the Trial of Galileo

Figure 1.1 is a picture of me talking about Renaissance astronomy to one of my classes at Texas A&M University. I like to wear this obvious costume once a semester just to get them in the mood. But this lecture might be the most important lecture of the semester. Why? We should not take intellectual freedom for granted. There are places in the world today where teachers do not have it. And in the Renaissance it was certainly not a given.

The philosopher Giordano Bruno (1546–1600) was charged with heresy by the Roman Inquisition. Not only did he advocate that the Earth revolves around the Sun. He suggested that the universe has countless other inhabited planets. We now know for certain that the Sun is the center of the solar system, and we have found hundreds of extra-solar planets. (We do not know if any harbor life.) Four hundred years ago Bruno was burned at the stake for expressing ideas that now seem quite sensible.

Galileo (1564–1642) had his own problems with the Roman Inquisition. These began after he had become famous for being the first astronomer to investigate celestial objects with a telescope, starting in 1610. Galileo discovered that the Moon had craters and mountains, Jupiter had four star-like objects revolving around it, and the telescope revealed more stars in the band of the Milky Way that one could see without a telescope.[1] Galileo also discovered that the Sun had spots and that Venus exhibited all the phases of the Moon (crescent, quarter, gibbous, full). Prior to that, Ptolemy's theory of the motion of Venus stipulated that it would always be situated between the Sun and the Earth, which implied that we would never see a gibbous or full phase.

In 1616 Galileo was summoned to Rome for an audience with Cardinal Robert Bellarmine. He was not officially in trouble, but he was informed that the notion that the Earth moved was inconsistent with Church doctrine, and might be heretical. He was advised to work on other things. At this time Copernicus' book *On the Revolutions of the Heavenly Spheres* was placed on the Index of Prohibited Books, as were other works that advocated the Earth's motion.

If one happened to have a copy of Copernicus' book, there were specific redactions that must be made.

In 1624 Cardinal Maffeo Barberini, a liberal intellectual, became Pope Urban VIII. Galileo went to Rome to wish his friend well, and also to ask permission to write a book that would discuss the arguments for and against the motion of the Earth. Then he spent eight years working on *The Dialogue Concerning the Two Chief World Systems*. The pros and cons of the Earth-centered (Ptolemaic) model of the solar system and the Sun-centered (Copernican) one were debated by one advocate of each, with an objective referee guiding the debate. The manuscript was submitted to the censors of the Inquisition, who demanded only minor revisions. Galileo's book was published in 1632.

What happened next has never clearly been resolved.[2,3] Pope Urban VIII was angry with Galileo because he felt Galileo had obtained permission in a devious manner to write his book. To Galileo's understanding the injunction of 1616 said that the motion of the Earth could not be set forth as "the truth." But then, in 1632, a scrutiny of the original document in the Vatican archives indicated that the motion of the Earth could not be discussed *in any way whatsoever*. Galileo's new book clearly discussed the motion of the Earth, so he was ordered to appear in front of the Roman Inquisition.

At his trial Galileo was forced to recant any belief in the motion of the Earth about the Sun. He was sentenced to house arrest and returned home to Arcetri, near Florence, where he stayed until his death in 1642.

Giorgio de Santillana makes a strong case in his book *The Crime of Galileo* that Galileo's trial was based on a judicial forgery,[4] namely that the key phrase "in any way whatsoever" was slipped into the injunction after 1616 and that Galileo knew nothing about it when he wrote his *Dialogue*. Three of the ten Cardinals judging Galileo refused to sign the sentence.[5]

In fact, Galileo had not proven that the Earth turns on its axis or revolves around the Sun. He thought that the tides proved the rotation of the Earth. They do not. Galileo did prove that not all bodies have to orbit the Earth; the Moons of Jupiter clearly revolve around Jupiter. He did not prove that Venus orbited the Sun, only that Ptolemy's model of the motion of Venus was incorrect.

The first *bona fide* evidence of the motion of the Earth was James Bradley's discovery in 1725 of the aberration of light.[6] It has to do with the speed of the Earth orbiting the Sun and the finite speed of light. The positions of stars are shifted in the direction of the Earth's motion by up to 20.5 arc seconds. The definitive proof that the Earth orbits the Sun was achieved via the first determinations of stellar parallax in the 1830's. The nearby stars move back and forth slightly (less than one arc second) on a yearly basis with respect to distant background stars owing to their being viewed from opposite ends of the diameter of the Earth's orbit.

Partly because of Bradley's discovery of aberration, in 1741 the Catholic Church allowed the publication of Galileo's collected works. In 1757 books advocating the Earth's motion were taken off the Index of Prohibited Books. By 1822 the Copernican paradigm could be presented as a thesis, not just a mere hypothesis.

Still, the whole Galileo affair festered in the minds of intellectuals for many generations. So, in 1981, by order of Pope John Paul II, a Pontifical Commission was established to look into the Ptolemaic-Copernican controversy. Their work was concluded in 1992. A summary of their findings is as follows:[7]

> The philosophical and theological qualifications wrongly granted to the then new theories about the centrality of the Sun and the movement of the earth were the result of a *transitional situation* in the field of astronomical knowledge, and of an exegetical *confusion* regarding cosmology....
>
> *It is in that historical and cultural framework, far removed from our own times, that Galileo's judges, incapable of dissociating faith from an age-old cosmology, believed, quite wrongly, that the adoption of the Copernican revolution, in fact not yet definitively proven, was such as to undermine Catholic tradition, and that it was their duty to forbid its being taught. This subjective error of judgment, so clear to us today, led them to a disciplinary measure from which Galileo "had much to suffer". These mistakes must be frankly recognized....*

Thus, 359 years after the trial of Galileo, the Catholic Church admitted it was a mistake to have put him on trial.

Today the Catholic Church is a strong supporter of scientific research. And as bizarre as the Big Bang theory is in some of its details, the Church seems to like the idea that the universe arose as a consequence of a single creation event. It is a bit like the opening verses of Genesis.

Any clash between science and religion in the United States seems to happen with Fundamentalist Protestants. Many reject the paradigm of natural selection for the evolution of life on Earth. Many discount radioactive dating methods as highly unreliable. To that I say, it depends on whether you are talking about absolute errors or relative errors. If a rock is determined to be 1 billion years old plus or minus 50 million years, to me that is a 5 percent relative error. Not bad. How many of us could estimate the age of a stranger to 5 percent if that person were considerably older or younger than we are?

I live in Texas, and every 10 years the State School Board sets new standards for textbook in schools. The nation watches closely what happens in Texas, because the textbook business is big business. Some members of the school board and their supporters take a strong anti-science stance. They say that we have to "teach the controversy." But there *is* no controversy about the basic ideas of evolution on the part of biologists. The controversy was made by Fundamentalist Christians. As I implied in Chapter 3 of this book, God could have made the universe any way He chose, but it is our job to determine how He actually made it. It is time for science texts to evolve.[8]

I am reassured by some recent comments by TV preacher Reverend Pat Robertson, that we must admit scientists are honestly trying to figure out things about the world; that the Earth is older than 6,000 years; and that dinosaurs roamed the Earth long before there were humans here.[9]

We live on a humble planet that revolves around a run-of-the-mill star in the outer reaches of the Milky Way galaxy. We can investigate our planet and can study light from the universe as far back as 400,000 years after the Big Bang, some 13.7 billion years ago. This realization should make us humble, and in our humility we should be thankful that we have the mental faculties to pose some daunting questions, and answer many of them.

Endnotes

1. Galilei, Galileo, *Sidereus Nuncius, or The Sidereal Messenger*, translted by Albert Van Helden, Chicago and London: University of Chicago Press, 1989.
2. Gingerich, Owen, "The Galileo Affair," *Scientific American*, August, 1982, pp. 132–143.
3. de Santillana, Giorgio, *The Crime of Galileo*, Chicago: Univ. of Chicago Press, 1955.
4. Ibid., pp. 283–297.
5. Ibid., p. 327.
6. Krisciunas, Kevin *Astronomical Centers of the World*, Cambridge: Cambridge University Press, 1988, pp. 86–87.
7. Address of His Eminence Cardinal Paul Poupard at the conclusion of the Proceedings of the Pontifical Study Commission on the Ptolemaic-Copernican Controversy in the 16th and 17th Centuries, 31 October 1992. Copy sent to Kevin Krisciunas on December 3, 1992, by Father George Coyne, former Director of the Vatican Observatory.
8. Wetherington, Ronald, and Egan, Scott P., "It's time for science texts to evolve," *Houston Chronicle*, November 17, 2013, p. B9.
9. http://www.huffingtonpost.com/2012/11/28/pat-robertson-creationism-earth-is-not-6000-years-old_n_2207275.html (accessed November 20, 2013).

Index

1984 (Orwell), 86

Abbot, Charles Greeley, 31
Abbott, Tim, 107–108
aberration of light, 116
Abetti, Giorgio, 31
Aborigines of Australia, 112
Aeschylus, 97
aether theory, 24
After Tamerlane: the Rise and Fall of Global Empires, 1400–2000 (Darwin), 92
AIDS. *See* human immunodeficiency virus
Aitken, Robert, 24
Ali and Nino (Nussimbaum), 110–111
Allende, Salvador, 111–112
Allen, Woody, 100
Almagest (Ptolemy), 37, 49
al-Shatir, Ibn, 37
Alvarez, Julia, 91
American Astronomical Society, 23, 34
American black lingo, 17
American Institute of Physics, 67
American Journal of Physics, 37

The Angel's Game (Zafón), 96
Anna Karenina (Tolstoy), 83
anthropology, 108–112
Apocalypse Now, 92
Aquinas, St. Thomas, 14
Arabic language, 17
Arbesman, Samuel, 3, 97
Aristarchus, 37
Aristophanes, 16
Asociación de Academias de la Lengua Española, 17
asteroids, 12
astronomers, 29–35
The Astronomers (TV series), 107
astronomical imaging, 43–46. *See also* telescopes
Astronomical Journal, 23, 24
astronomical measurements, 37–42. *See also* Earth; Moon; Sun
astronomical standard candles, 41–42
"The Awful German Language" (Twain), 16

Bacon, Kevin, 106–107
Barabási, László, 106

119

Barberini, Cardinal Maffeo, 117
Barnacle, Nora, 100
Baronius, Caesar Cardinal, 14
Basel problem, 59
Beach, Sylvia, 100
Beg, Ulugh, 30
Bellarmine, Cardinal Robert, 115
bell-shaped curve, 61–62
Benedictine Abbey of Fulda, 96
Bey, Mohammad Essad. *See* Nussimbaum, Lev
Bible, 14, 74
Biggest Elvis (Kluge), 87
Black Death, 76–77
Blake, Robert, 103–104
Blink (Gladwell), 106
Bloom, Leopold, 100, 101
Blumenthal, Walter Hart, 26
Boltzmann, Ludwig, 30
Bond, George Phillips, 44
The Book of Lost Books (Kelly), 97–98
Book Lust (Pearl), 83, 101
books
    about books
        The Book of Lost Books (Kelly), 97–98
        Book Lust (Pearl), 83, 101
        Great Books (Denby), 1
        More Book Lust (Pearl), 83
    bound in human skin, 25–26
    first lines in, 83–87
    history of, 1
    lost books, 95–98
    most brilliant thesis, 21–22
    most egotistical, 22–25
    prohibited, 115, 116
    with a moral angle, 89–93
bottleneck effect, 76
Bracciolini, Poggio, 96
Bradley, James, 116

Brahe, Tycho, 32
Brief Biography and Popular Account of the Unparalled Discoveries of T. J. J. See, 22, 23
The Brief Wondrous Life of Oscar Wao (Diaz), 91
Bruno, Giordano, 30, 115
Bubka, Sergei, 11
Budgen, Frank, 101
Bulgakov, Mikhail, 85
Bulletin of the American Astronomical Society, 32
Burton, Richard Francis, 15
Buss, David, 108

Cabanès, Dr., 26
Caesar, Julius, (note 4), 71
calculus, 53–64
The Call of the Wild (London), 86
Calvino, Italo, 98, 99
Campbell, William Wallace, 30
Card, Orson Scott, 86
Catholic Church, 115–117
cats, 75
Cepheids (stars), 41–42
Chamberlain, Richard, 107
Chandrasekhar limit, 67
Chandrasekhar, Subrahmanyan, 67
charge coupled devices (CCDs), 44, 46
Charlois, Auguste, 30
Chatwin, Bruce, 112
cheating, 79–81, 105
Chilean language, 17–18
Chilean National Observatory, 30
Christakis, Nicholas, 107
CIA, 111
Citizen Vince (Walter), 87
clickers, 80–81
Clinton administration, 111–112
Coca-Cola, 73

Collapse (Diamond), 111
The Columbia History of the World, 1
Connected (Christakis and Fowler), 107
Conrad, Joseph, 92
constellations, 49–50
constellation song, 50
Cook's Branch Nature Conservancy, 67
Copernicus, Nicholas, 115–117
Coppola, Francis Ford, 92
cosine, 56–57
Cosmicomics (Calvino), 99
cosmological distance ladder, 40–41
The Crime of Galileo (Santillana), 116
Crime and Punishment (Dostoyevsky), 90
crime statistics, 105
Crossley telescope, 44, 45
cultures, 109–112

Darconville's Cat (Theroux), 101, 102
Dark Matter (film), 30
Darwinism, 108
Darwin, John, 92
Dead Souls (Gogol), 97–98
death, causes of, 29, 30
Délimitation Scientifique des Constellations (Delporte), 49–50
Delporte, Eugène, 49–50
Denby, David, 1
Derbyshire, John, 93
The Dialogue Concerning the Two Chief World Systems (Galileo), 116
Diamond, Jared, 92, 109, 111
Diaz, Junot, 91
Dickens, Charles, 83
distance equation, 9
dogs, 75
Dostoyevsky, Fyodor, 90, 102
Doyle, Sir Arthur Conan, 85–86
Draper, Henry, 44, 45
The Dreams of Scipio (Pears), 89–90

Drunken Goldfish and Other Irrelevant Scientific Research (Hartson), 73
Dubner, Stephen, 79–80, 105
Dunbar's Rule, 106

Earth
  age of, 14
  axial tilt, 41
  circumference, 39
  eccentricity of orbit, 40
  motion of, 116
  obliquity of the ecliptic, 39
  radius, 39
Educational Testing Service (Princeton, NJ), 53
Einstein, Albert, 13, 24, 77
elemental abundance, 21
Empire's End: A History of the Far East from High Colonialism to Hong Kong (Keay), 92
Encyclopedia Brittanica, 2–3
Encyclopédie, 1
energy of motion. *See* kinetic energy equation
English language, 15, 16, 17, 101
eponyms, 68–70
equilateral polygons, 55–56
Erdös Number, 106
Erdös, Paul, 06
Escalante, Jaime, 53
ethics, 79–81
Euler, Leonhard, 59–61
Euripides, 97
evolution, 74
exponential growth rates, 3
extinct societies, 111

Facebook, 106
fare-beating, 106
Faulkner, William, 103
Fear of Flying (Jong), 86

The Feast of the Goat (Llosa), 90–91
feminists, 108
first lines, in books, 83–87
Fitch, Noel Riley, 100
Fitzgerald, F. Scott, 86, 91, 100
Flammarion, Camille, 25–26
formulas. *See* mathematical equations
Fourier analysis, 59–60
Fowler, James, 107
Fowles, John, 86
Freakonomics (Levitt and Dubner), 79–80
French language, 15, 17
frogs, 76
fruit flies, 74
Fundamentalist Protestants, 117

Galileo, 9, 43, 115–117
Gaposchkin, Sergei, 21–22
Gardner, Martin, 58
Gauss, Carl Friedrich, 61–64
Gaussian distribution, 63–64
gender ratios, in athletes, 5–8
General Relativity theory, 24
genetics, 74, 76–77
Gerasimovich, Boris Petrovich, 30
German language, 16, 18
Gerson, Levi ben, 37
Gingerich, Owen, 26
The Girls of Atomic City: The Untold Story of the Women Who Helped Win World War II (Kiernan), 112
Gladwell, Malcolm, 105–106
gnomon, 38–39, 40
The God of Animals (Kyle), 87
Goertz, Christoph, 30
Gogol, Nikolai, 97–98
goldfish, 73
Google Earth, 39
graffiti, 105–106
Great Books (Denby), 1

The Great Gatsby (Fitzgerald), 91
Greenblatt, Stephen, 96
Grossman, Leonid, 102
Guns, Germs, and Steel (Diamond), 92, 109, 111
guppies, 74

Hale, George Ellery, 22, 23
The Half-Life of Facts (Arbesman), 3, 97
Hardy, G. H., 58
Hardy-Ramanujan number, 58
Harper, William Rainey, 23
Hartson, William, 73
harumphing, 110
Harvard College Observatory, 44
Harvard University, 22, 73
Harvey, Dr. Thomas Stoltz, 77
Hawaiian language, 18
Hawking, Stephen, 67
Heart of Darkness (Conrad), 92
Hemingway, Ernest, 100, 104
Hermite, Charles, 58
Herschel, William, 43
high-rise syndrome, 75
Hipparchus, 49
History of Julius Caesar (Napoleon III), 90
Hoffleit, Dorrit, 21–22, 31–32
Holmes, Sherlock, 102
human immunodeficiency virus (HIV), 77, 79, 81
Hypatia, 30

Index of Prohibited Books, 115, 116
integral calculus, 53–56
intellectual freedom, 115
intelligent design, 74
Internet Movie Database, 106–107
In the Time of the Butterflies (Alvarez), 91

inverse function, 56–57
Ironman competition, 8
Isinbayeva, Yelena, 11
Italian language, 15, 17, 18

Jacobsen, Theodore, 32
James Joyce and the Making of Ulysses and other Writings (Budgen), 101
Japanese language, 16
Jong, Erica, 86
Joyce, James, 100–101

Kafka, Franz, 83
Keay, John, 92
Keeler, James, 44, 45
Kelly, Stuart, 97–98
Kepler's Third Law of planetary motion, 41
Khan, Genghis, 74
Kiernan, Denise, 112
kinetic energy equation, 9, 12
King Solomon, 74
Kissinger, Henry, 111
Kluge, P. F., 87
Kornbluh, Peter, 111
Kyle, Aryn, 87

Lacaille, Nicolas Louis de, 49
La Disparition (Perec), 98
Lahaye, Thierry, 40
Landes, David, 109–110
language, 15–18. *See also individual languages*
Larsdatter, Live, 32
latitude, measuring, 39
Leech, John, 58
Lehrer, Tom, 50
Leibniz, Gottfried Wilhelm, 59
Lenin's Tomb: The Last Days of the Soviet Empire (Remnick), 112
Letelier, Orlando, 111
Levitt, Steven, 79–80, 105

libraries, 110–111
Lick Observatory, 24, 30, 34, 44, 45
life expectancy, 29, 31, 32, 33, 34, 107
Linked (Barabási), 106
Link, Goethe, 31
Lithuanian language, 17
Llosa, Mario Vargas, 90–91
London, Jack, 86
longest word, 16
longitude, measuring, 39
Lorenz, Konrad, 75–76
Lowell Observatory, 22, 23, 24
Love in the Time of Cholera (Marquez), 84
Lucretius, 96–97
Lu, Gang, 30

Macaulay, Rose, 83
The Magus (Fowles), 86
"The Major General's Song" (*The Pirates of Penzance*), 50
The Making of the Atomic Bomb (Rhodes), 112
Manhattan Project, 112
Marks, Rodney, 30
marriage, 107
Marquez, Gabriel Garcia, 83–84
The Master and Margarita (Bulgakov), 85
mathematical equations
    area under the curve, 53–55, 61–63
    center of mass, 11, 12
    distance, 9
    Gaussian distribution, 63
    kinetic energy, 9, 12
    normal curve of error, 63
    potential energy, 9
    spectral radiance, 60
    total energy, 10
    velocity, 9, 10
    volume of a sphere, 12
mating rituals, 74, 76
Matthew Effect, 106

Mays Business School, 80
McCutcheon, Robert, 30
McDonald Observatory, 24–25
Melville, Herman, 84–85
Menocal, María Rosa, 110
Merton, Robert K., 68
Messier 81 (galaxy), 45
The Metamorphosis (Kafka), 83
Midnight in Paris (movie), 100
Moby Dick (Melville), 84–85
Moneyball (movie), 107
Mongolian language, 17
monogamy, 108
Monteiro, Hektor, 108
Moon
    angular diameter, 38
    angular size of, 37, 38, 39
    distance to, 37
    eccentricity value, 38
    lunar eclipse, 37
    model of motion of, 37
    perigee to perigee period, 38
    sighting device, 38
The Moral Animal (Wright), 76, 107–109
morality, 89–93, 109
More Book Lust (Pearl), 83
Morita, Koichiro, 30
Mt. Wilson Observatory, 44
murder rates, 30

Napoleon III, 90
National Institute of Health and Medicine, 77
National Public Radio, 83
Newman, James R., 57–58
Newton, Isaac, 59, 64
New York City, 105–106
New Yorker (magazine), 105
New York Times (newspaper), 107
Nicholson, Dwight, 30

Nicoll, James, 15
Nixon, President Richard, 111–112
Novaya Gazeta (magazine), 102
nuclear weapons, 12, 13
numerical integration, 54
Nussimbaum, Lev, 110–111
The Nutmeg of Consolation (O'Brian), 86–87
Nyad, Diana, 7

O'Brian, Patrick, 86–87
O'Brien, Stephen, 76–77
obscure words, 101
observatories
    Chilean National Observatory, 30
    Harvard College Observatory, 44
    Lick Observatory, 24, 30, 34, 44, 45
    Lowell Observatory, 22, 23, 24
    McDonald Observatory, 24–25
    Mt. Wilson Observatory, 44
    US Naval Observatory, 22
The Offshore Pirate (Fitzgerald), 86
Olympics, 5, 6, 7
On Aggression (Lorenz), 75–76
One Hundred Years of Solitude (Marquez), 83–84
On the Nature of Things (Lucretius), 96–97
On the Revolutions of the Heavenly Spheres (Copernicus), 115–117
Operation Condor, 111
The Orientalist (Reiss), 111
Orion Nebula, 44, 45
Ornament of the World (Menocal), 110
Orwell, George, 86
OuLiPo (l'Ouvroir de Littératre Potentialle), 98, 99
Outliers (Gladwell), 106

Palomar telescope, 46
Parseval's equation, 60

Payne(-Gaposchkin), Cecilia, 21–22
PBS, 107
Pearl, Nancy, 83, 101
Pears, Iain, 89–90
Pepsi-Cola, 73
Perec, George, 98
Perlmutter, Saul, (note 4), 71
"Pile On" (Blake), 103–104
Pinochet, Augusto, 111–112
The Pinochet File (Kornbluh), 111
The Pirates of Penzance (Gilbert and Sullivan), 50
Poisson statistics, 79
pole vaulting, 10–12
Pope Alexander VI, 2
Pope Gregory XIII, (note 2), 71
Pope John XXIII, 96
Pope John Paul II, 117
Pope Urban VIII, 116
potential energy equation, 9
Prime Obsession: Bernhard Riemann and the Greatest Unsolved Problem in Mathematics (Derbyshire), 93
The Prisoner of Heaven (Zafón), 96
Protestants, Fundamentalist, 117
proton-proton cycle, 13
Ptolemaic-Copernican controversy, 117
Ptolemy, Claudius, 37, 49, 115
Pylon (Faulkner), 103

quantum efficiency (QE), of photographic emulsions, 44, 46
Quechua language, 17

Ramanujan's constant, 58
Ramanujan, Srinivasa, 57–58
rats, 76
Ravaud, Dr., 25–26
Real Academia Española, 17
Regis, Ed, 77

Reiss, Tom, 111
Remnick, David, 112
Rhodes, Richard, 112
Ridpath, Ian, 49–50
Rieman zeta function, 60
Riess, Adam G., (note 4), 71
Ristenpart, Friedrich Wilhelm, 30
Robertson, Reverend Pat, 117
Robinson, Sir George, 110
Roman Inquisition, 115
roundoff error, 55
running
  Olympics, 5, 6, 7
  pole vaulter, 10
  sprinter, 10
  world records, 5, 6
Russell, Henry Norris, 21
Russian language, 17, 18
Ryback, R. S., 73

Sagan, Carl, 2
Said, Kurban. *See* Nussimbaum, Lev
San Francisco Journal, 24
Santillana, Giorgio de, 116
same-sex marriage, 107
Savonarola, Girolamo, 2
Schaffhauser, General, 111–112
Schindler, Oskar, 92
Schindler's List (Schindler), 92
Schmerer, Helen, 2
Schmidt, Brian P., (note 4), 71
Schönberg-Chandrasekhar limit, 67
Schwarz, Hugo, 108, 109–110
Schwarzschild, Karl, 34
Scientific American (magazine), 58
Seeing Calvin Coolidge in a Dream (Derbyshire), 93
See, Thomas Jefferson Jackson, 22–25
The Shadow of the Wind (Zafón), 95–96

Shakespeare and Company (Paris bookstore), 100
Shan, Linhua, 30
Sharif, Ismail ibn, "Moulay Ismail the Bloodthirsty," 74–75
shortest story possible, 104
The Sidereal Messenger (Galileo), 43
The Signal and the Noise (Silver), 8, 107
Sign of Four (Doyle), 85–86
Silver, Nate, 8, 107
sine, 56–57
Sirius (star), 21
Six Degrees of Kevin Bacon, 106–107
Sky and Telescope (magazine), 29–30, 31, 34
Smith, Robert A., 30
snakes, 76
social networks, 106, 107
Sollenberger, Paul, 32
The Song Lines (Chatwin), 112
Sophocles, 97
The Sound and the Fury (Faulkner), 103
Soviet Union, 112
Spanish history, 110–111
Spanish language, 17, 18
Speaker for the Dead (Card), 86
sperm, 73, 74
Spielberg, Steven, 92
sprinters, 10–11
Stand and Deliver (movie), 53
stars, atmospheres of, 21
Star Trek, 17
statistics, 29–30, 79–81
Stein, Gertrude, 100
Stellar Atmospheres (Payne), 21
Stevenson, Robert Louis, 85
Strakhov, Nikolai Nikolaevich, 90
Struve, Otto, 34
subway system (NYC), 105–106
suicide rates, 30
Summer in Baden-Baden (Tsypkin), 102–103

Sun, 13–14, 21
  maximum elevation angle, 40
  obliquity of the ecliptic, 40
  total energy emitted, 13, 61
Super Freakonomics (Levitt and Dubner), 80
supernova 2012fr, 46
supernovae, Type Ia (stars), 42
surveys, 79–81
Swahili language, 17
The Swerve: How the World Became Modern (Greenblatt), 96–97
Swift satellite, 46
swimming
  long distance, 7
  Olympics, 7
Sylvia Beach and the Lost Generation (Fitch), 100

A Tale of Two Cities (Dickens), 83
tangent, 55–56
taxicab numbers, 58
The Tears of the Cheetah (O'Brien), 76–77
telescopes
  collecting area, 43
  Crossley reflector, 44, 45
  focal length, 43
  light-gathering power, 43
  Palomar, 46
  reflecting, 44, 45, 46
  refracting, 43–44
Ten Thousand Hour Rule, 106
Terres du Ciel (Flammarion), 26
Texas A&M University, 80–81, 115
Thackeray, William Makepeace, 85
Theroux, Alexander, 101, 102
Tikopia, 111
The Tipping Point (Gladwell), 105–106
Tolstoy, Leo, 25, 83, 84, 90
The Towers of Trebizond (Macaulay), 83
transcendental numbers, 58

Transylvania, 106
Treasure Island (Stevenson), 85
trigonometric parallax, 41
trigonometry, 56–57
Tsar Bomba, 12, 13
Tsypkin, Leonid, 102–103
turkeys, 76
Twain, Mark, 16

Ulysses (Joyce), 100–101
University of Chicago, 22, 23, 67
University of Iowa, 30
University of Padua, 74
Urquhart, Sir Thomas, 97
US Library of Congress, 1
US Naval Observatory, 22

Vanity Fair (Thackeray), 85
Vega (star), 21
velocity equation, 9, 10, 11
Villari, Linda, 3
volume of a sphere, 12

Walker, Dick, 23–24
Walter, Jess, 87
Warren Harding effect, 106
The Wealth and Poverty of Nations (Landes), 109–110
Webb, W. L. (William Larkin), 22–25
Who Got Einstein's Office: Eccentricity and Genius at the Institute for Advanced Study (Regis), 77
The World of Mathematics (Newman), 57–58
world records, running, 5, 6
Wright, Robert, 76, 107–109
writing, 15
writing styles, 99–104

Yiddish language, 17

Zafón, Carlos Ruiz, 95–96
Zulu language, 18